Cambridge O Level
Physics
With Stafford

Complete notes and worked examples

Syllabus code: 5054

Compiled

by

Dr. Stafford Valentine Redden

(B.Sc. (Hons.); B.Ed.; M.Sc.; M.Ed.; M.A.; Ph.D)
Founder
www.staffordeducationalservices.com

ISBN: 978 81 910705 6 9

DEDICATION

This book is dedicated to www.staffordeducationalservices.com

Contents

Chapter One
Physical Quantities, Units and Measurement

Cambridge 5054 syllabus specification 1(a) define the terms scalar and vector.
Cambridge 5054 syllabus specification 1 (b) determine the resultant of two vectors by a graphical method.
Cambridge 5054 syllabus specification 1 (c) list the vectors and scalars from distance, displacement, length, speed, velocity, time, acceleration, mass and force.
Cambridge 5054 syllabus specification 1 (d) describe how to measure a variety of lengths with appropriate accuracy using tapes, rules, micrometers and calipers. (The use of a vernier scale is not required.)
Cambridge 5054 syllabus specification 1 (e) describe how to measure a variety of time intervals using clocks and stopwatches.
Cambridge 5054 syllabus specification 1 (f) recognise and use the conventions and symbols contained in 'Signs, Symbols and Systematics', Association for Science Education, 2000.

1 (a, c) Scalar and vector quantities

The physical quantities that we often measure can be classified as scalar and vector quantities.

Scalar	Vector
Scalar quantities have magnitude (size) but **no direction**.	Vector quantities **have** magnitude (size) and **have direction**.
For example a distance of 5m.	For example a force of 25N acting 15^0 North East.
Magnitude 5m Unit	Magnitude 25N acting towards 15^0 North East Unit Direction
Likewise for a body of mass 10Kg. 10 is the magnitude and Kg is the unit.	Likewise for a car being displaced by 15m North. 15 is the magnitude, m is the unit and the direction of displacement is towards North.
Some examples of **scalar** quantities Mass, Distance, Speed, Time	Some examples of **vector** quantities Force or weight, Displacement, Acceleration, Velocity

1 (d) Measuring length

In real life situation, we often have to measure the length of many different objects. Accurate measurements will require the use of appropriate equipment. The table below gives the list of instruments that can be used to measure specific lengths with appropriate accuracy and precision.

Length to be measured	Suitable instrument	Accuracy of instrument
Several metres (m)	Measuring Tape	0.1 cm
Several centimetres (cm) to 1m	Ruler	0.1 cm
Between 1 cm to 10 cm	Vernier calipers	0.1mm or 0.01 cm
Less than 2 cm	Micrometer screw gauge	0.01mm or 0.001 cm

Examples of appropriate use of instruments

The length of a cricket pitch can be measured accurately with a measuring tape, shown in figure 1.1.

The length of a notebook can be measured appropriately by using a ruler, as shown in figure 1.2.

Figure 1.1: Measuring tape

Figure 1.2: Ruler

Using a Vernier calipers

A Vernier calipers is used to accurately measure the inner or outer diameter of a test tube or rings. It can also be used to measure the diameter of a sphere. The diagram of a Vernier calipers is shown in figure 1.3.

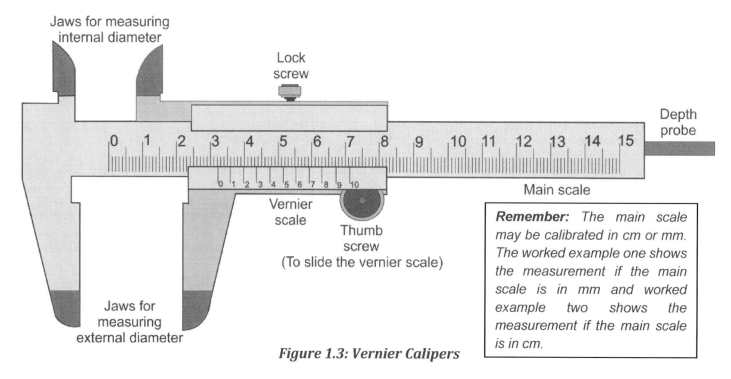

Figure 1.3: Vernier Calipers

Remember: *The main scale may be calibrated in cm or mm. The worked example one shows the measurement if the main scale is in mm and worked example two shows the measurement if the main scale is in cm.*

Worked example one

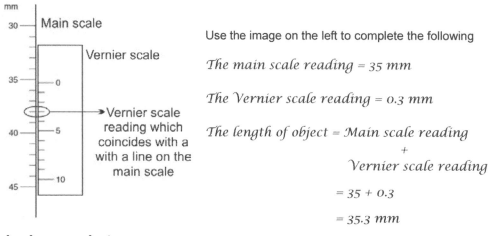

Use the image on the left to complete the following

The main scale reading = 35 mm

The Vernier scale reading = 0.3 mm

The length of object = Main scale reading
+
Vernier scale reading

= 35 + 0.3

= 35.3 mm

Remember:
Least count in mm = 0.1mm
Vernier Scale Reading (VSR) = Vernier division which coincides with main scale x least count

So, VSR = 3 x 0.1
= 0.3 mm

Worked example two

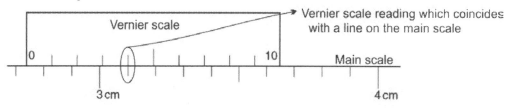

Use the image above to complete the following
The main scale reading (MSR) = 2.7 cm
The vernier scale reading (VSR) = 0.04 cm
The length of object = MSR + VSR
= 2.7 + 0.04
= 2.74 cm

Least count in cm = 0.01cm
So, VSR = 4 x 0.01
= 0.04 cm

Using a micrometer screw gauge

A micrometer screw gauge measure the diameter of circular object with a high precision of 0.01mm. The screw gauge is shown in figure 1.4.

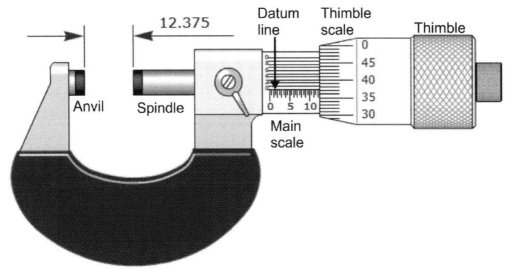

Figure 1.4: Micrometer screw gauge

Worked example three

Use the image on the left to complete the following

Main scale reading (MSR) = 15mm

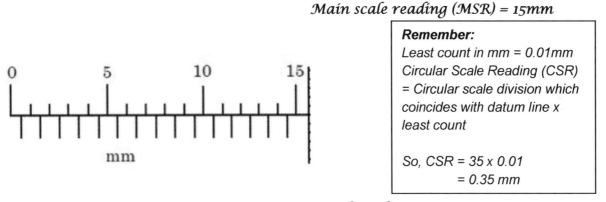

> **Remember:**
> *Least count in mm = 0.01mm*
> *Circular Scale Reading (CSR)*
> *= Circular scale division which coincides with datum line x least count*
>
> *So, CSR = 35 x 0.01*
> *= 0.35 mm*

Final reading = MSR + CSR = 15.35 mm

1 (e) Measuring time

Time can be measured in years, months, weeks, days, hours, minutes, seconds, milliseconds, etc. The SI unit for measuring time is seconds (s). A variety of clocks and stopwatches are used for measuring time. The clock with the highest precision is the atomic clock and the clock with the lowest precision is the pendulum clock.

Type of clock	Use and accuracy
Atomic clock	10^{-10} s
Digital stopwatch	10^{-1} s
Analogue stopwatch	1 s
Ticker-tape timer	0.02 s
Pendulum clock	min

Ticker timers

Ticker timers are usually used to measure the speed or acceleration of trolleys moving along a bench. The arrangement is shown in figure 1.5.

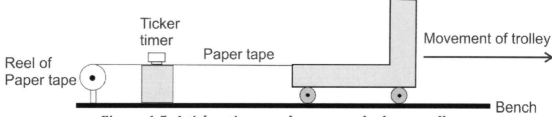

Figure 1.5: A ticker timer and tape attached to a trolley

Most ticker-timers vibrate at 50 Hz, and thus make 50 dots per second. For these, the expected mean value of one tick is 1/50 second, or 0.02 s.

The time represented by the ticker tape shown in figure 1.6 is calculated to be 0.62 s.

Figure 1.6: A used ticker tape

Working
Number of gaps = 31
Time period of ticker = 0.02 s
Time = 31 x 0.02 = 0.62 s

1 (b) Resultant force of two vectors (Graphical method)

The resultant of two vectors can be determined by the graphical method. Vector A and Vector B act as shown in the figure 1.7. The resultant of the two vectors can be measured by the length of the diagonal of the parallelogram, as shown in the figure. The direction of the resultant vector is θ North East with respect to the horizontal vector B.

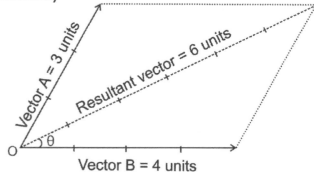

Figure 1.7: Resultant force of vectors

Activity

Two forces A and B act upon a body as shown by the arrows A and B. The arrow E shows the resultant force on the body. The diagram below is drawn to the scale where 1 cm represents 1 N force.

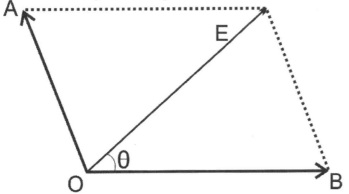

Use a ruler to measure the following.

The force A =

The force B =

The resultant force E =

Use a protractor to measure angle θ =

State the direction and magnitude of the resultant force E.

Chapter Two
Kinematics

Distance	Displacement
The total length of the path travelled by a body. Irrespective of direction.	The shortest length between the initial and final position of a body.
Is a **scalar** quantity. It is not measured in a specific direction.	Is a **vector** quantity. It is measured in a specific direction.
Measured in m or cm.	Measured in m or cm.

2 (a, b) Speed

Speed is the distance travelled by a body per unit time. It is a **scalar** quantity (it has no specific direction). Figure 2.1 shows a graph for a body moving with **uniform speed** of 10 m/s.

Velocity

Velocity is the distance travelled in a **specific direction** (displacement) per unit time. It is a **vector** quantity. Figure 2.2 shows a graph for a body moving with a **uniform velocity** of 10 m/s and **zero acceleration**.

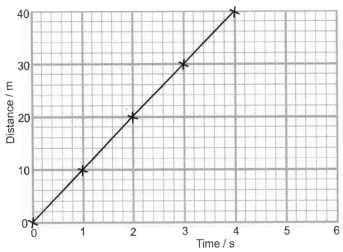

Figure 2.2: Graph showing Uniform speed

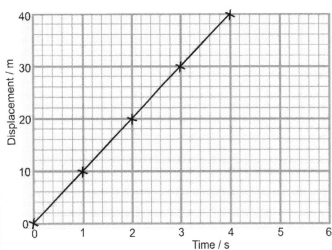

Figure 2.2: Graph showing Uniform velocity

$$\text{Speed} = \frac{distance(s)}{\text{time(t)}}$$

$$\text{Average speed} = \frac{\text{Total distance travelled}(s)}{\text{Total time taken (t)}}$$

$$\text{Velocity} = \frac{displacement(s)}{\text{time(t)}}$$

Figure 2.3 shows a graph for a body moving with **increasing velocity**.

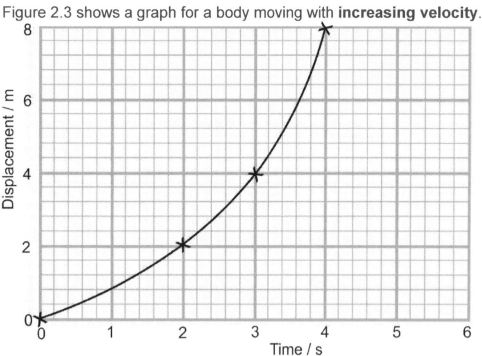

Figure 2.3: Graph showing increasing velocity

Remember:

A body moving with increasing velocity will cover greater distance in a specific direction per unit time.

For example, the average velocity during the first 2 seconds is
Velocity = displacement / time
= 2 / 2
= 1 m/s

However, the average velocity between 2 and 4 seconds is
Velocity = displacement / time
= 6 / 2
= 3 m/s
The velocity has increased from 1 m/s to 3 m/s.

2 (c) Acceleration

Acceleration is the rate of change of velocity of a body. It is a **vector** quantity.

$$\text{Acceleration} = \frac{Change\ in\ velocity}{time\ taken\ (t)} \quad \text{or} \quad \text{Acceleration} = \frac{Final\ velocity\ (v) - Initial\ velocity\ (u)}{time\ taken\ (t)}$$

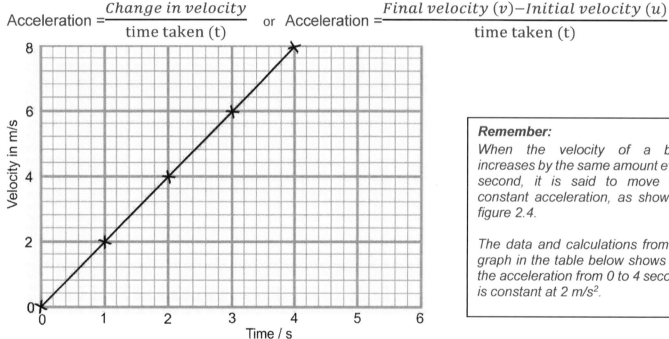

Figure 2.4: Graph showing constant acceleration

Remember:
When the velocity of a body increases by the same amount every second, it is said to move with constant acceleration, as shown in figure 2.4.

The data and calculations from the graph in the table below shows that the acceleration from 0 to 4 seconds is constant at 2 m/s².

Time Interval / s	Initial velocity (u) in m/s	Final velocity (v) in m/s	Acceleration = (v − u) /t in m/s²
0 to 1	0	2.0	Working: (2.0 − 0) / 1 = **2.0**
1 to 2	2.0	4.0	Working: (4.0 − 2.0) / 1 = **2.0**
2 to 3	4.0	6.0	Working: (6.0 − 4.0) / 1 = **2.0**
3 to 4	6.0	8.0	Working: (8.0 − 6.0) / 1 = **2.0**
Notice that the change in velocity is the same with every second.			

2 (d) Non-uniform acceleration

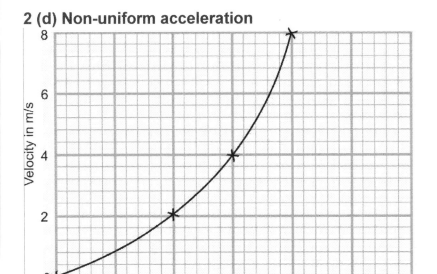

Figure 2.5: Graph showing increasing acceleration

Remember:

When the change in velocity of a body increases with every second, it is said to move with increasing acceleration, as shown in figure 2.5.

The data and calculations from the graph in the table below shows that the acceleration from 0.8 m/s² in the 1ˢᵗ second to 4 m/s² between the 3ʳᵈ and 4ᵗʰ second.

Time Interval / s	Initial velocity (u) in m/s	Final velocity (v) in m/s	Acceleration = (v – u) /t in m/s²
0 to 1	0	0.8	Working: (0.8 – 0) / 1 = **0.8**
1 to 2	0.8	2.0	Working: (2.0 – 0.8) / 1= **1.2**
2 to 3	2.0	4.0	Working: (4.0 – 2.0) / 1= **2.0**
3 to 4	4.0	8.0	Working: (8.0 – 4.0) / 1= **4.0**
Notice that the change in velocity increases with every second.			

2 (e) Interpretation of velocity-time graphs

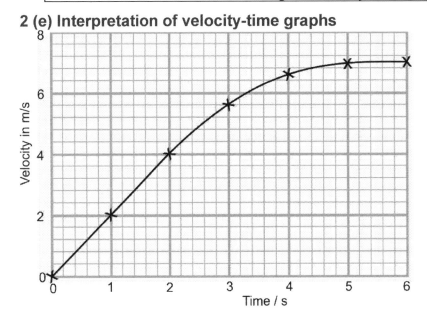

Figure 2.6: Graph showing change in velocity over time

Remember:

The gradient of a velocity-time graph gives the acceleration.

Interpretation:

- *Between 0 to 2 seconds, the body moves with constant acceleration, velocity increases.*
- *Between 2 to 5 seconds, the body moves with a decreasing acceleration, but velocity still increases.*
- *Between 5 and 6 seconds, the body moves with a constant velocity and zero acceleration.*

Worked example

A car moves from A to B for 5m and then travels a further 3m to reach point C. Find the distance and displacement of the car.

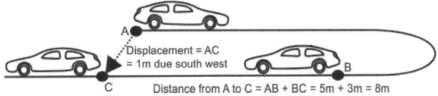

Distance = 5 + 3 = 8m

Displacement = 1m towards South west (A to C)

Interpretation of displacement-time graphs

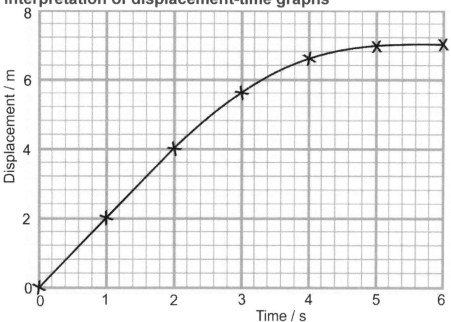

Figure 2.7: Graph showing change in displacement over time

Remember:
The gradient of a displacement - time graph gives velocity.

Interpretation:

- *Between 0 to 2 seconds, the body moves with uniform velocity, acceleration is zero.*

- *Between 2 to 5 seconds, the body moves with a decreasing velocity, acceleration is negative.*

- *Between 5 and 6 seconds, the body is at rest, velocity is zero.*

2 (f, g) Calculating distance travelled in a velocity-time graph

The area under a velocity-time graph is used to determine the distance travelled for motion with uniform velocity or uniform acceleration, as shown in figure 2.8.

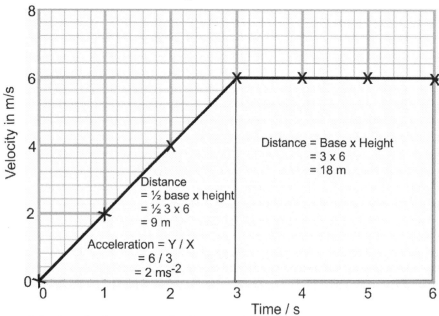

Figure 2.8: Graph showing calculation of distance in a velocity=time graph

Total distance moved in 6 seconds = 9+18

= 27 m

Remember:

Problems on bodies moving with **CONSTANT acceleration** can be solved using the equations of motion. There are 4 equations:

$v = u + at$

$s = \frac{1}{2}(u+v)t$

$s = ut + \frac{1}{2}at^2$

$v^2 - u^2 = 2as$

Where, u = initial velocity; a = acceleration; v = final velocity; s = displacement; t = time taken.

The known values of s, u, v, a and t, can be substituted in the appropriate formula to find the unknown quantity.

2 (h, i) Free-fall for a body near to the Earth

When an object above the ground is released from rest, it fall vertically and accelerates at 10 ms^{-2}, when there is no air resistance.

All objects, irrespective of their mass, will fall freely with the same acceleration. So, a feather and a stone will fall at the same acceleration in a vacuum.

The speed of the object increases uniformly by 10 ms^{-1} each second, as shown in figure 2.9.

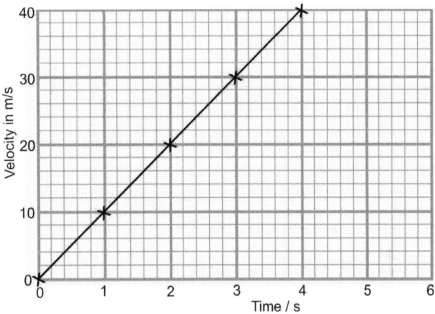

Figure 2.9: Graph showing constant acceleration during free fall without air resistance

When an object falls through air its acceleration decreases to zero due to air resistance. It then falls at a constant velocity known as terminal velocity, as shown in figure 2.10. During terminal velocity the air resistance is equal to the weight of the body.

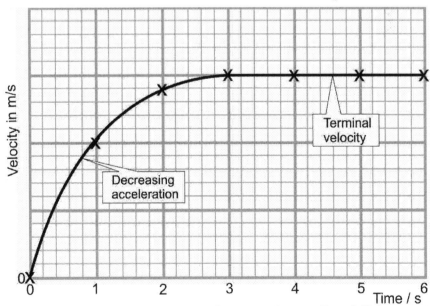

Figure 2.10: Graph showing change in acceleration during free fall due to air resistance

Chapter Three
Forces

Cambridge 5054 syllabus specification 3 (a) state Newton's third law.

Cambridge 5054 syllabus specification 3 (b) describe the effect of balanced and unbalanced forces on a body.

Cambridge 5054 syllabus specification 3 (c) describe the ways in which a force may change the motion of a body.

Cambridge 5054 syllabus specification 3 (d) recall and use the equation force = mass × acceleration.

Cambridge 5054 syllabus specification 3 (e) explain the effects of friction on the motion of a body.

Cambridge 5054 syllabus specification 3 (f) discuss the effect of friction on the motion of a vehicle in the context of tyre surface, road conditions (including skidding), braking force, braking distance, thinking distance and stopping distance.

Cambridge 5054 syllabus specification 3 (g) describe qualitatively motion in a circular path due to a constant perpendicular force, including electrostatic forces on an electron in an atom and gravitational forces on a satellite. (F = mv 2/r is not required.)

Cambridge 5054 syllabus specification 3 (h) discuss how ideas of circular motion are related to the motion of planets in the solar system.

A force is a pull or push on a body. It has both magnitude and direction, so it is a **vector**. Force is measured in "Newton" (N).

Forces can change:
- the speed of an object
- the direction that an object is moving in
- the shape of an object.

4 (b) Newton's First law of motion

A body at rest will remain at rest until an unbalancing force is exerted upon it. This tendency of the body to remain at rest is called as **inertia of rest**.

 A toy car on a horizontal surface will remain at rest until an external force is applied on it.

Figure 3.1: Inertia of rest

> **Remember:**
> A larger force will have to be applied to move a heavy vehicle from its position of rest than to move a lighter vehicle.

A body in motion will continue to move until a force is exerted upon it to oppose the motion. This is called as **inertia of motion**. Friction is usually an opposing force which brings moving objects to rest.

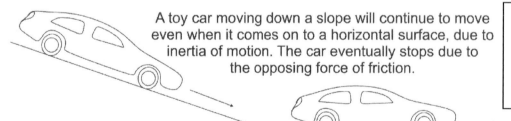 A toy car moving down a slope will continue to move even when it comes on to a horizontal surface, due to inertia of motion. The car eventually stops due to the opposing force of friction.

> **Remember:**
> If the mass of the body is high then inertia will be high. So, a larger force will be needed to stop a heavy vehicle than a lighter vehicle.

Figure 3.2: Inertia of motion

3 (c, d) Newton's Second Law

The rate of change of momentum is proportionate to the resultant force acting on a body.

It tells us how much an object accelerates (a) if a force (F) acts on a body of mass (m). The force is equal to the product of mass and acceleration of the body.

Force = mass x acceleration

or

F = m x a

Worked example one	**Worked example two**
A student of mass 55 kg sit on a bicycle of mass 30kg. If the bicycle is pushed, and accelerates at 4 m/s², how big is the force pushing it? *Answer: First, we need to add the mass of the student and the bicycle together. That's 55kg + 30kg = 85kg we know that acceleration = 4 m/s² Using F = ma, we have F = 85 x 4 = 340 N*	A feather of mass 2g falls in a vacuum towards the earth. A force of 0.02 N acts on the feather. Calculate the acceleration of the feather.

Worked example two (continued):

Given,   *Using the equation,*
$m = 2 / 1000$ $a = F/m$
    $= 0.002 \ kg$ $= 0.02 / 0.002$
$F = 0.02 N$ $= 10 \ ms^{-2}$

3 (a) Newton's third law
Newton's third law of motion states that every action has an equal and opposite reaction.

For example, when you walk, you push against the ground with your feet. The ground then exerts an equal and opposite force and pushes you forward. Likewise, a book resting on a table exerts a downward force on the table top and the table top exerts an equal and opposite upward (reaction) force to balance the book. Also, when a bullet is fired from a gun, the bullet exerts a backward force on the gun, which is called as the recoil.

3 (b) Balanced and unbalanced forces on a body.
When several forces act on an object, we can add them together to find out what's going to happen. The combined effect of all the forces on an object is called the resultant force.

Example 1: Two forces act upon the wooden block as shown below. The resultant force is 5N towards the east.

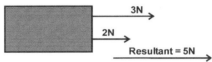

Figure 3.3: Forces acting on a body in the same direction

Example 2: Two forces act upon the wooden block as shown below. The resultant force is 1N towards the west.

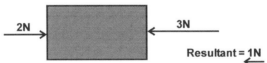

Figure 3.4: Forces acting on a body in opposite directions

Example 3: The weight of the car (W) is balanced by an equal and opposite force called the resultant force (R) exerted by the road on the car.

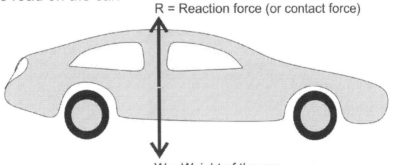

R = Reaction force (or contact force)

W = Weight of the car
(Acts downwards due to the pull of gravity)
Figure 3.5: Vertical forces acting on a car at rest

Worked example
If you push a car with a force of 500N, and somebody else pushes the other way with a force of 400N, what is the resultant force on the car?
Answer: *500 – 400 = 100 N in the direction of the larger force*

3 (e) Friction and motion

Friction is a force which opposes motion of a body. It occurs when two surfaces rub against each other. Friction provides the force to accelerate, stop or change the direction of a body.

Friction is caused by:

- Chemical bonds between molecules in the two surfaces.
- Electrostatic forces of attraction between the two surfaces.
- Physical barriers (roughness)

Air resistance is also a type of friction. The motion of a parachutist can be used to demonstrate the effect of friction on the motion of a body, as shown in figure 3.6.

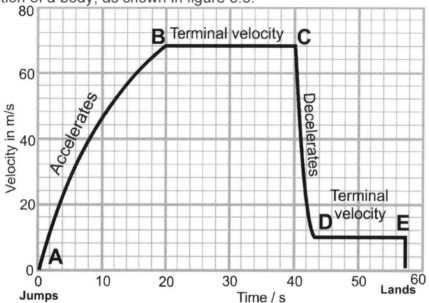

Figure 3.6: Graph showing the change in velocity of a parachutist over time

The table below describes and explains the motion of the parachutist during free fall.

A to B The parachutist has just jumped out of the aircraft, and is falling faster and faster *("accelerating" – velocity increases)*. As his **velocity increases**, the **air resistance also increases**. Resultant force on the parachutist at the moment when he jumped from the aircraft is 700N acting downwards and acceleration is maximum. The acceleration will decrease as the air resistance increases.	700 N
B to C The parachutist is now falling fast enough for the **air resistance** to **equal** his **weight**. The resultant force is zero, so acceleration is zero. This means that the **forces** on him are **in balance**, so his **velocity** stops increasing and **stays constant** - he has reached his **terminal velocity**.	700 N 700 N
C to D As the parachute opens, the air resistance (friction) increases dramatically, so the parachutist slows down greatly *("decelerates")*. For a few seconds, the air resistance is greater than his weight. As he decelerates, the air resistance decreases.	1100 N 700 N
D to E The parachutist has slowed down, and now the **forces are back in balance**. His **velocity** is **constant** again - he has a **new**, much slower, **terminal velocity**.	700 N 700 N

3 (f) Thinking distance

The **thinking distance** is the **distance travelled by the vehicle** as the driver makes a decision to apply the brakes. It is the time needed for the reaction to occur after seeing the stimulus.

Thinking distance can be increased by:
- Greater speed – if the vehicle is travelling at a high speed, then the vehicle will travel further while the driver is thinking about applying the brakes.

- Alcohol – alcohol slows down the transmission of nerve impulses. So, the reaction time increases and so does the thinking distance.

Braking distance

The braking distance is the distance travelled by the vehicle after the brakes are applied.

Braking distance can be increased by:
- Higher speed – Higher speed will increase the braking distance.
- Poor road conditions – wet roads and icy roads will reduce friction between the tyres and the road and increase braking distance.
- Vehicle conditions – wide tyres will increase the surface in contact with the road and decrease the braking distance. Poor brakes will mean that braking distance will increase. If the vehicle is full of people or loaded with more mass then the braking distance will increase.

Stopping distances

Under normal driving conditions, thinking distance and braking distance depend on the speed of the vehicle.
The stopping distance is much **further** for **faster** speeds, which is why you should:
- Keep your **distance** from the vehicle in front, especially if the road conditions are poor.
- Keep to the speed limit.

The figure 3.7 shows the motion of a car from the moment the driver sees an obstacle until the car is brought to a stop.

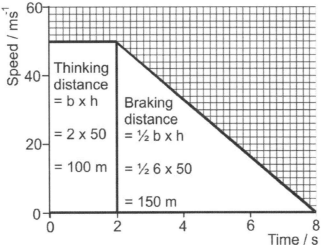

Figure 3.7: Graph showing thinking distance and braking distance

The driver sees an obstacle at time 0 seconds.
The driver applies the brakes at 2 seconds.
The car comes to a halt at 8 seconds.
Stopping distance = thinking distance + braking distance
$$= 100 + 150$$
$$= 250 \, m$$

3 (g) Circular motion

For a body to move in a circular path, a centripetal force (F1) acts towards the center of the circular path. The outward force (F2) is the direction in which the body will move if the centripetal force is removed.

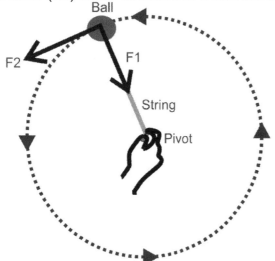

Figure 3.8: Circular motion and centripetal force

In circular motion, the centripetal force (F1) acts towards the center of the circular path. This force keeps the body moving in a circular path, as shown in figure 3.8.

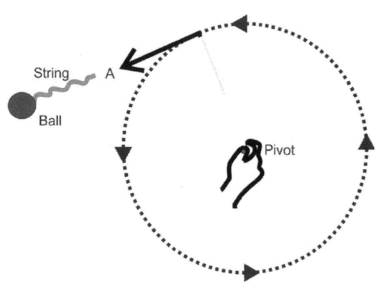

Figure 3.9: Path followed by a body if the centripetal force is removed

If the centripetal force is removed, then the body will stop moving in a circular direction and will move away in the direction A, as shown in figure 3.9.

3 (h) Solar system and centripetal forces

Gravitational force which keeps planets in circular motion is centripetal. **Gravitational force depends on the mass of the object and the distance from the centre**. Sun has the biggest mass in our solar system. Therefore, the gravitational force strength of Sun is the highest and this keeps all planets towards itself. Only larger objects keep smaller objects around them. Larger objects pull smaller objects towards them. This gravitational force is centripetal that is why circular motion is possible, as shown in figure 3.10.

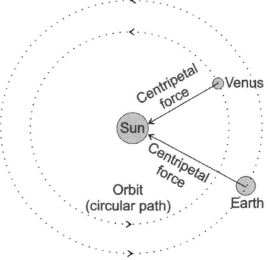

Figure 3.10: Revolution of planets in the solar system

Cambridge 5054 syllabus specification 4 (a) state that mass is a measure of the amount of substance in a body.
Cambridge 5054 syllabus specification 4 (b) state that the mass of a body resists change from its state of rest or motion.
Cambridge 5054 syllabus specification 4 (c) state that a gravitational field is a region in which a mass experiences a force due to gravitational attraction.
Cambridge 5054 syllabus specification 4 (d) recall and use the equation weight = mass × gravitational field strength.
Cambridge 5054 syllabus specification 4 (e) explain that weights, and therefore masses, may be compared using a balance.
Cambridge 5054 syllabus specification 4 (f) describe how to measure mass and weight by using appropriate balances.
Cambridge 5054 syllabus specification 4 (g) describe how to use a measuring cylinder to measure the volume of a liquid or solid.
Cambridge 5054 syllabus specification 4 (h) describe how to determine the density of a liquid, of a regularly shaped solid and of an irregularly shaped solid which sinks in water (volume by displacement).
Cambridge 5054 syllabus specification 4 (i) define density and recall and use the formula density = mass/volume.

4 (a, e) Mass (m)

Mass is the amount of matter contained in a body. The mass does not change from one place to another. Mass is measured by an electronic balance. The unit is g or kg.

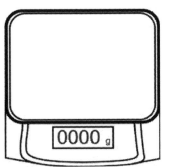

> **Remember:**
> *To measure the mass of very small objects like pins, measure the mass of a certain number of pins (50 for example) and then divide the mass of the pins by the number of pins. This helps to reduce the measurement error and makes the measurement more accurate.*
>
> *Accuracy is the closeness of the measured value to the real value.*

Figure 4.1: Electronic top pan balance

4 (c, d, f) Weight (W)

Gravity is a force which is exerted on all bodies in the gravitational field. The weight of the body is the product of mass and gravitational field strength. It can be measured by using a spring balance. Weight changes from one place to another according to the strength of gravity of the place. Weight is measured in Newton (N).

> **Remember:**
> **Weight (N) = mass (kg) x gravitational field strength (N/kg)**
>
> *The weight of a body changes with the gravitational field strength, however the mass remains constant.*
> *The gravitational field strength on Earth is 10 N/kg.*
> *The gravitational field strength on the moon is 1.6 N/kg.*
> *So, if a body of mass 9kg has a weight of 90N on the Earth, the weight of the same object on the moon will be*
> $$W = m \times g$$
> $$= 9 \times 1.6$$
> $$= 14.4 \, N$$

Figure 4.2: Spring balance

4 (g) Measuring volume of a liquid

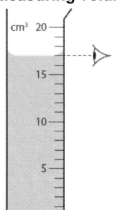

Remember:
The volume refers to the space occupied by an object. The volume of a liquid can be measured by a measuring cylinder, as shown in figure 4.3. The unit for volume is cm³ or m³.

*The level of the eye should be kept horizontally at the same level as the lower meniscus of the liquid in the measuring cylinder to avoid parallax errors. This results in more **accurate** measurement of volume.*

The volume of liquid in the measuring cylinder in figure 4.3 is 17 cm³.

Figure 4.3: Measuring cylinder

Measuring volume of an irregular solid

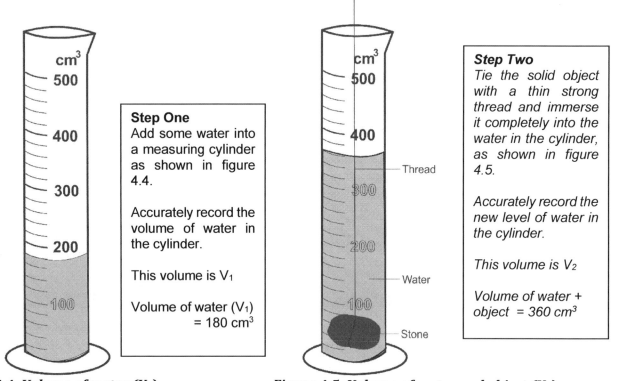

Step One
Add some water into a measuring cylinder as shown in figure 4.4.

Accurately record the volume of water in the cylinder.

This volume is V_1

Volume of water (V_1) = 180 cm³

Step Two
Tie the solid object with a thin strong thread and immerse it completely into the water in the cylinder, as shown in figure 4.5.

Accurately record the new level of water in the cylinder.

This volume is V_2

Volume of water + object = 360 cm³

Figure 4.4: Volume of water (V_1)   *Figure 4.5: Volume of water and object (V_2)*

Step three
Volume of the object (stone) = $V_2 - V_1$
$$= 360 - 180$$
$$= 180 cm^3$$

4 (h) Density (ρ)
Density is the mass per unit volume of a body. It is calculated by dividing the mass with the volume of an object. The unit is g cm⁻³ or kg m⁻³. Density is a scalar quantity.

For example, if the mass of the stone measured in figure 4.5 is 50g, then the density of the stone could be found as follows.

Density (ρ) = $\dfrac{Mass\ (m)}{Volume\ (V)}$
= 50 / 180
= 0.28 g cm⁻³

Remember:
It is better to measure the mass before immersing the stone in water, otherwise the stone will be wet and the mass cannot be accurately measured.

4 (i) Worked examples

Calculate the density of a piece of wood measuring 30 cm x 20 cm x 5 cm and of mass 2.50 kg. Give your answer

a) in gcm⁻³	b) in kgm⁻³.
$Volume = lxbxh$ $\quad = 30x20x5$ $\quad = 3000 \; cm^3$ $Mass = 2.5 \; x \; 1000$ (Remember: 1kg = 1000g) $\quad = 2500 \; g$ Density (ρ) $= \dfrac{Mass \; (m)}{Volume \; (V)}$ $\quad = 2500 \; / \; 3000$ $\quad = 0.83 \; g \; cm^{-3}$	$Volume = lxbxh$ $\quad = 0.3x0.2x0.05$ (Remember: 1m = 100 cm) $\quad = 0.003 \; m^3$ $Mass = 2.5$ $\quad = 2.5 \; kg$ Density (ρ) $= \dfrac{Mass \; (m)}{Volume \; (V)}$ $\quad = 2.5 \; / \; 0.003$ $\quad = 833 \; kg \; m^{-3}$

c. A sheet of copper is used to make the roof of a building. The copper sheet has dimensions 4.0 m x 3.2 m x 0.008m. The density of copper is 8940 kgm⁻³. Calculate the mass of the copper sheet.

$m = ρ \; x \; V$
$\quad = 8940 \; x \; (4x3.2x0.008)$
$\quad = 915 \; kg$

Density of an object which is too large to fit into a measuring cylinder.
First the Eureka can is filled with water to the level of the spout, as shown in figure 4.6.
The mass of a solid can then be measured by an electronic top pan balance or beam balance.
The solid is tied with a thread and immersed into a displacement (Eureka) can, as shown in figure 4.7.
The water flowing out of the spout is collected into a measuring cylinder and the volume of the solid is measured.

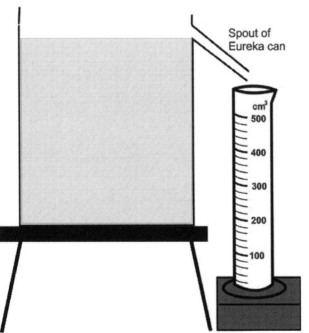

Figure 4.6: Fill the Eureka can with water

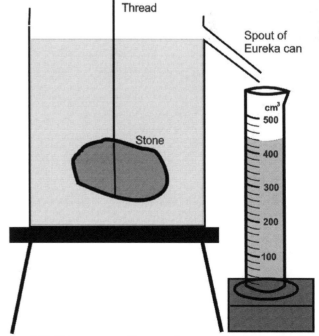

Figure 4.7: Measure volume of water in the cylinder

Chapter Five
Turning Effect of Forces

> *Cambridge 5054 syllabus specification 5* (a) *describe the moment of a force in terms of its turning effect and relate this to everyday examples.*
> *Cambridge 5054 syllabus specification 5* (b) *state the principle of moments for a body in equilibrium.*
> *Cambridge 5054 syllabus specification 5* (c) *define moment of a force and recall and use the formula moment = force × perpendicular distance from the pivot and the principle of moments.*
> *Cambridge 5054 syllabus specification 5* (d) *describe how to verify the principle of moments.*
> *Cambridge 5054 syllabus specification 5* (e) *describe how to determine the position of the centre of mass of a plane lamina.*
> *Cambridge 5054 syllabus specification 5* (f) *describe qualitatively the effect of the position of the centre of mass on the stability of simple objects.*

5 (a) Moment

Moment of a force is the turning effect of a force. The unit of moment of a force is Nm. It is a vector quantity. The turning effect of a force about a fixed point (pivot) is shown in figure 5.1 and 5.2.

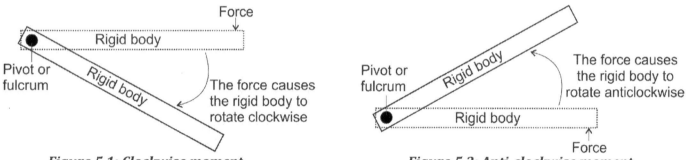

Figure 5.1: Clockwise moment   *Figure 5.2: Anti-clockwise moment*

> **Remember:**
> Moment (M) = Force (F) x perpendicular distance of the force from the pivot (d)

5 (b) Principle of moments

The principle of moments states that when the body is in equilibrium, the total clockwise moment is equal to the total anticlockwise moment.

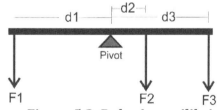

Figure 5.3: Ruler in equilibrium

> **Remember:**
>
> At equilibrium,
> Clockwise moment = Anticlockwise moment
>
> (F2 x d2) + (F3 x d3) = F1 x d1

5 (c, d) Worked Example one

A force of 5 Newton acts in the direction of F1 in figure 5.3 and the distance d1 is 22cm. Another force of 3N acts in the direction F2 at a distance of 15cm (d2) from the pivot. Calculate the force F3 needed to balance the ruler if F3 acts at 32cm from the pivot.

At equilibrium, *Clockwise moment = Anticlockwise moment* (F2 x d2) + (F3 x d3) = F1 x d1 (3 x 15) + (X x 32) = (5 x 22)	$45 + 32X = 110$ $32X = 110 - 45$ $32X = 65$ $X = 65 / 32$ $X = 2.03$ N

Cambridge GCE O Level Physics with Stafford   Page 18

Worked Example two

The machine below is used to punch holes in aluminum sheets. The force applied on the lever is very small and provides a mechanical advantage. Two forces act on either side of the punch. The first force is the 5 N force acting upward when the lever is pressed. The punch is just about to puncture a hole in the aluminum.

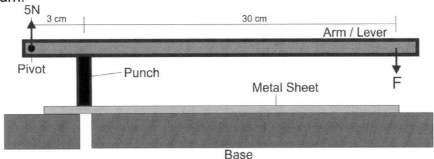

At equilibrium,
The anticlockwise moment = 5 x 3

$$= 15 \; Ncm^{-1}$$

The clockwise moment = F x 30

$$= 30 \; F \; Ncm^{-1}$$

Since,
The anticlockwise moment = The clockwise moment

$$15 = 30 \; F$$
$$15 / 30 = F$$
$$F = 0.5 \; N$$

The advantage of using this machine is that a small force can be used to punch a hole in the aluminum sheet.

Remember:
The pivot here is the punch.

The clockwise moment is exerted by the force F and acts at a distance of 30 cm from the punch.

The anticlockwise moment is exerted by a 5N force acting 3 cm from the punch.

Worked Example three

A person of 80 kg mass walks on a plank of negligible weight. The plank is supported by two trestles. The distance between the trestles X and Y is 10m. The forces A and B act on the plank to support the weight of the person. The forces can be calculated as shown below.

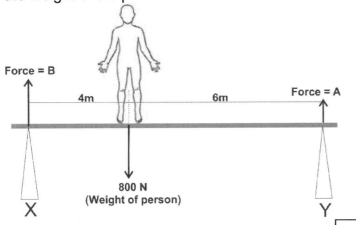

Remember:
Clockwise moment about X is 3200Nm.

The trestle Y is exerting an upward force to balance the clockwise moment about trestle X. This upward force contributes to the anticlockwise moment about trestle X.

So, anticlockwise moment about trestle X, is the product of distance from trestle X (10m) and the upward force (A) exerted by trestle Y on the plank.

Moment about X
Anticlockwise moment = Clockwise moment

$$A \times 10 = 800 \times 4$$
$$10A = 3200$$
$$A = 320 \; N$$

Force B = 800 - 320
$$= 480 \; N$$

Similarly, taking moment about Y
Anticlockwise moment = Clockwise moment

$$800 \times 6 = B \times 10$$
$$4800 = B \times 10$$
$$B = 480 \; N$$

Force A = 800 - 480
$$= 320 \; N$$

Worked Example four

A ruler of mass 15g is supported by a pivot as shown in the diagram below. The center of mass is shown by the point x. Use the information in the diagram to determine the distance at which the 25g mass has to be suspended to balance the ruler.

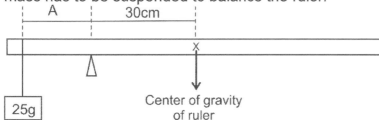

Center of gravity
of ruler

25g

At equilibrium,
Anticlockwise moment = Clockwise
moment
25 x A = 15 x 30
25A = 450
A = 450/25
A = 18cm

Centre of mass of simple objects

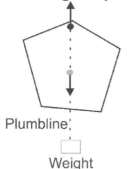

Remember:
The centre of mass is the point through which the whole weight of the body tends to act. The centre of mass of a few regular shaped objects is shown in figure 5.4.

Figure 5.4: Centre of mass of a few regular shaped objects

5 (e) Determining the position of the centre of mass of a plane lamina

Plumbline

Weight

Pin

Weight

Lamina rotates freely to allow the centre of gravity to lie below the pin

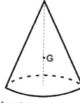

C

Step one: Use a pin to pivot the lamina. The lamina should be able to rotate freely about the pin.

Step two: Once the lamina comes to rest, suspend a plumb line from the pin and draw a line along the plumb line.

Step three: Repeat the procedure by pivoting the lamina at different points. The point at which the three lines intersect is the centre of mass.

5 (f) Stability of an object

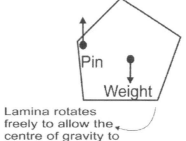

Figure 5.5: Destabilising a stable object

Remember:
An object is said to be stable if it is steady and well balanced so that when it is pushed slightly it does not topple or fall off easily.
An unstable object will easily topple over if it is tilted slightly.

Fact: Bodies with low center of gravity and a wide base are more stable than bodies with a high center of gravity and narrow base.

A - the weight of the body passes through the base, so the body will remain stable.
B - the box is tilted and then released. The box falls back on to its base, because the weight is still passing through the base.
C – the box is tilted and then released. The box falls on to its side, because the weight does not pass through the base anymore.

Chapter Six
Hooke's law

Cambridge 5054 syllabus specification 6 (a) *state that a force may produce a change in size and shape of a body.*
Cambridge 5054 syllabus specification 6 (b) **plot, draw and interpret extension-load graphs for an elastic solid and describe the associated experimental procedure.*
Cambridge 5054 syllabus specification 6 (c) **recognise the significance of the term "limit of proportionality" for an elastic solid.*
Cambridge 5054 syllabus specification 6 (d) *calculate extensions for an elastic solid using proportionality.*

6 (a, b) When we apply a force to a spring, it stretches.
Measure the original length of the spring when there is no force suspended from the spring. This is called the unstretched length.
Measure the length of the spring after hanging a weight (Load) from the spring. This is called as the stretched length.

Extension = Stretched length – Unstretched length

Hooke's Law states that within the elastic limit the extension is <u>directly proportional</u> to the stretching force (load). The spring will go back to its original length when the force is removed **as long as elastic limit (E) is not exceeded.**

If the spring is stretched beyond the elastic limit then it remains permanently stretched even after the stretching force is removed, as illustrated in figure 6.1.

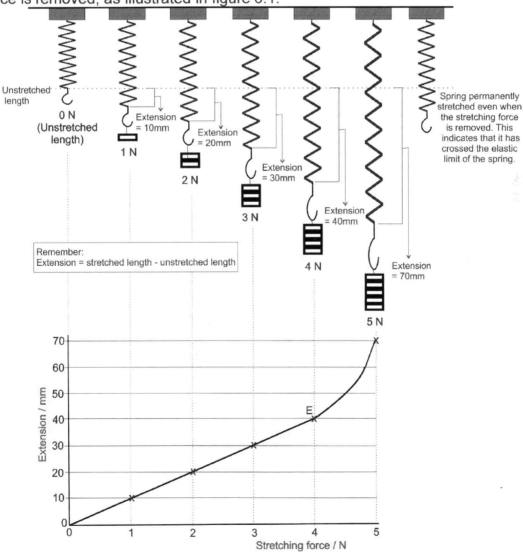

Figure 6.1: Demonstration of Hooke's law

6 (c) Below the elastic limit, we say that the spring is showing "elastic behaviour": the extension is proportional to the force, and it'll go back to it's original length when we remove the force.

Beyond the elastic limit, we say that it shows "plastic behaviour". This means that when a force is applied to deform the shape, it stays deformed when the force is removed.

We use Hooke's Law in spring balances, kitchen scales and other devices where we measure using a spring.

6 (d) Worked example

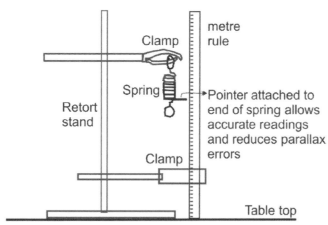

The length of the spring when no load is suspended from the spring is 15cm.

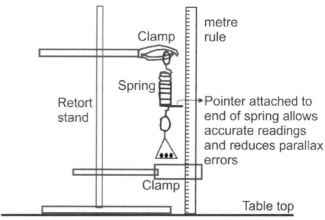

The length of the spring when a 8N load is suspended from the spring is 21cm.

Calculate the extension of the spring.
Extension = stretched length - unstretched length

$$= 21 - 15$$

$$= 6cm$$

Calculate the extension of the spring that you would expect if a load of 10N is suspended from the spring.

8N load -------------- extension is 6cm

(Cross multiply)

10N load -------------- X cm

$8X = 6 \times 10$

$X = (6 \times 10) / 8$

$$= 7.5 \ cm$$

State the stretched length of the spring when a load of 10N is acting upon it.

Extension = stretched length - unstretched length

7.5 = stretched length - 15

Stretched length = 7.5 + 15

$$= 22.5 \ cm$$

> *Cambridge 5054 syllabus specification 7 (a) define the term pressure in terms of force and area, and do calculations using the equation pressure = force/area.*
> *Cambridge 5054 syllabus specification 7 (b) explain how pressure varies with force and area in the context of everyday examples.*
> *Cambridge 5054 syllabus specification 7 (c) describe how the height of a liquid column may be used to measure the atmospheric pressure.*
> *Cambridge 5054 syllabus specification 7 (d) explain quantitatively how the pressure beneath a liquid surface changes with depth and density of the liquid in appropriate examples.*
> *Cambridge 5054 syllabus specification 7 (e) recall and use the equation for hydrostatic pressure $p = \rho gh$.*
> *Cambridge 5054 syllabus specification 7 (f) describe the use of a manometer in the measurement of pressure difference.*
> *Cambridge 5054 syllabus specification 7 (g) describe and explain the transmission of pressure in hydraulic systems with particular reference to the hydraulic press and hydraulic brakes on vehicles.*
> *Cambridge 5054 syllabus specification 7 (h) describe how a change in volume of a fixed mass of gas at constant temperature is caused by a change in pressure applied to the gas.*
> *Cambridge 5054 syllabus specification 7 (i) recall and use $p_1V_1 = p_2V_2$.*

7 (a, b) Pressure (P)

Pressure is the force acting per unit area. Pressure is measured in Pascals (Pa). Pressure is a vector quantity. A pressure of 1 Pascal means a force of 1 Newton is acts per square metre of area.

Pressure depends on two factors:
- the **Force** (in Newtons) and
- the **Area** on which the force is being exerted (in square metres)

$$\text{Pressure (P)} = \frac{Force\ (F)}{Area(A)}$$

Worked examples

A wooden block of mass 25kg has a length of 2m, height of 0.5m and breadth of 1m when resting on its largest surface. Calculate the force exerted by the block of wood on the floor. Show your working.

Weight (W) = Mass (m) x gravitational force (g)
= 25 x 10
= 250 N

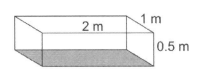

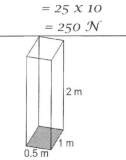

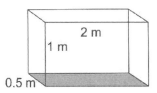

Calculate the area of the shaded area.

Area = l x b
= 2 x 1
= 2 m²

Calculate the pressure exerted by the block on the floor.

Pressure = Force / Area
= 250 / 2
= 125 Pa

Calculate the area of the shaded area.
Area = l x b
= 0.5 x 1
= 0.5 m²

Calculate the pressure exerted by the block on the floor.
Pressure=Force/Area
= 250 / 0.5
= 500 Pa

Calculate the area of the shaded area.

Area = l x b
= 2 x 0.5
= 1 m²

Calculate the pressure exerted by the block on the floor.

Pressure = Force / Area
= 250 / 1
= 250 Pa

A large force might only create a small pressure if it's spread out over a wide area. Also, a small force can create a big pressure if the area is tiny.

- When the area is small, a moderate force can create a very large pressure. This is why a sharp knife is good at cutting things: when you push the very small area of the sharp blade against something, it creates a really large pressure.

- Even a slender supermodel can damage to floors by using pencil-heeled shoes. This is because the area of the heel is small, that it can easily create enough pressure to cause a dent in the floor.

7 (c) Mercury Barometer

The piece of lab equipment specifically designed to measure the pressure of gases is known as the barometer. A **barometer** uses the height of a column of mercury to measure gas pressure in millimeters of mercury. The mercury is pushed up the tube from the dish until the pressure at the bottom of the tube (due to the mass of the mercury) is balanced by the atmospheric pressure, as illustrated in figure 7.1.

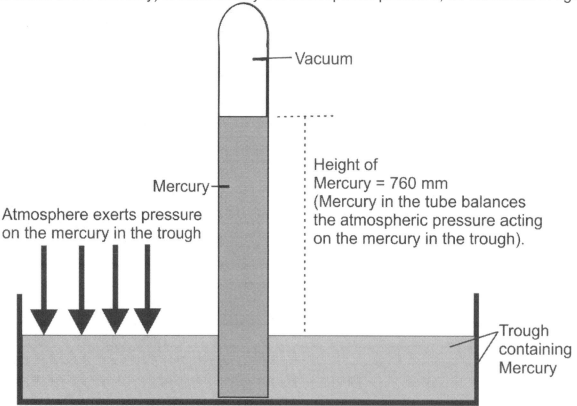

Figure 7.1: A simple mercury barometer

The atmospheric pressure is expressed as 76 cm of mercury or 760 mm of mercury.

The pressure in Pascal can be calculated by the equation

$Pressure = \rho g h$

$= 13600 \times 9.81 \times 0.76$

$= 101396\ Pa$

$= 101.4\ k\ Pa$

> **Remember:**
> Density (ρ) of mercury = 13600 kg m^{-3}
> g = 9.81 ms^{-2} (g is taken as 10 ms^{-2} in O level)
> h = 0.76 m
> mercury is used as a barometric liquid because of its high density. If water or alcohol were used instead of mercury, a very long tube would have to be used, which would be impractical.

7 (d, e) Pressure beneath a liquid surface changes

Pressure in liquids depends upon three factors

ρ - Density of the liquid

g – Acceleration due to gravity

h – depth from the surface of the liquid

Pressure = ρ x *g* x *h*

When designing the walls of pools and dams, the wall must be **thicker** at the bottom, to withstand the increased pressure down there. This is because the pressure increases with depth, as shown in figure 7.2.

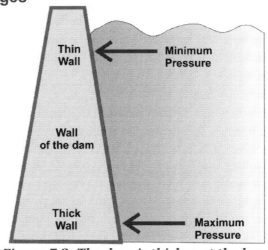

Figure 7.2: The dam is thicker at the base

7 (f) Describe the use of a manometer in the measurement of pressure difference

A U-tube manometer is shown in figure 7.3. The manometer liquid level is the same in both sides of the tube, as atmospheric pressure is acting equally on both the surfaces of the liquid.

Atmospheric pressure is the same on both surfaces of the liquid, so liquid level in both stems of the manometer is the same

Atmospheric pressure is the same on both surfaces of the liquid, so liquid level in both stems of the manometer is the same

> **Remember:**
> Pressure exerted by liquid column = ρgh
> Pressure exerted by gas = Atmospheric pressure + liquid pressure

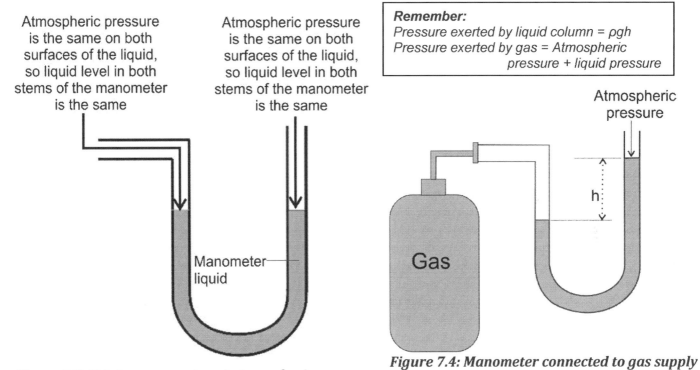

Figure 7.3: U-tube manometer at atmospheric pressure

Figure 7.4: Manometer connected to gas supply

Worked example

Use the information in figure 7.4 to measure the gas pressure. The liquid used in the manometer has a density of 13600 kg m^{-3} and the difference in height (h) of the liquid levels in both stems of the tube is 30cm. The atmospheric pressure is 101 kPa.

Pressure exerted by liquid column = ρ x *g* x *h*
$$= 13600 \times 10 \times 0.30$$
$$= 40,800 \ Pa \ or \ 40.8 \ kPa$$
Pressure exerted by gas = *Atmospheric pressure* + *liquid pressure*
$$= 101 + 40.8$$
$$= 141.8 \ kPa$$

7 (g) Hydraulic systems

Hydraulic systems work by applying force to one piston, which then transmits the force to the second piston through oil in a pipe. Since, oil is incompressible almost all the applied force is transmitted to the second piston. The force is usually multiplied by increasing the area of the second piston, as illustrated in figure 7.5.

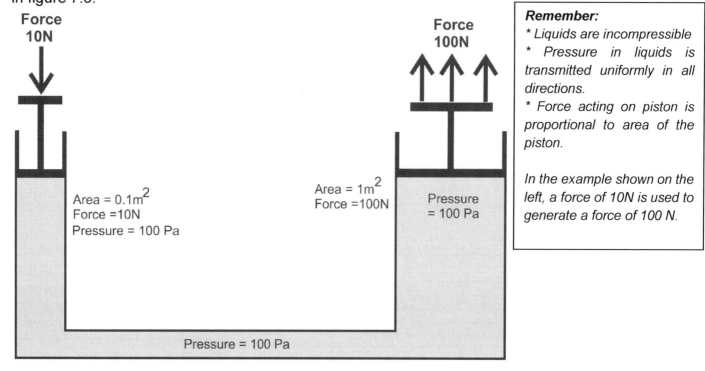

Remember:
* Liquids are incompressible
* Pressure in liquids is transmitted uniformly in all directions.
* Force acting on piston is proportional to area of the piston.

In the example shown on the left, a force of 10N is used to generate a force of 100 N.

Figure 7.5: Hydraulic system

Hydraulic press

The hydraulic press is used to lift a huge load by applying a small effort. Hydraulic systems are used in diggers, car brakes and fairground rides. A pump creates pressure in an incompressible liquid, which acts on a piston. By adjusting the area of the piston, we can adjust the force we get.

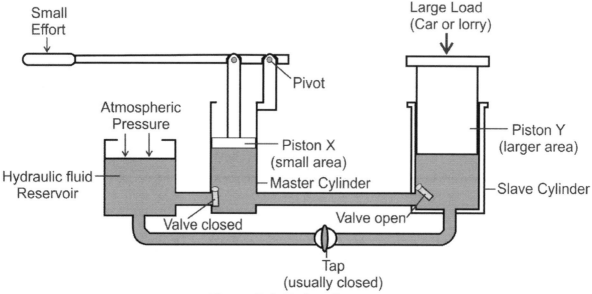

Figure 7.6: Hydraulic press

When the piston X is pushed down, it exerts a pressure on the hydraulic fluid. This pressure is transmitted to piston Y. the large area of piston Y ensures that a large force is generated to lift a lorry or car, as shown in figure 7.6.

Worked example

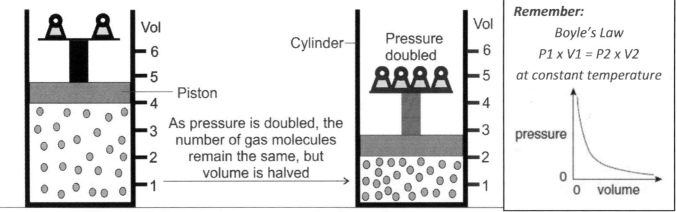

The diagram shows the hydraulic braking system of a car. A small force applied on the brake pedal exerts a large force on the brake shoes.

The master cylinder has an area of 12 cm² and a force of 60N is applied on the master cylinder.

i. Calculate the pressure on the master cylinder.

$$P = F / A$$
$$= 60 / 12$$
$$= 5 N cm^{-2}$$

ii. The slave cylinder has an area of 30cm². Calculate the force exerted on the slave cylinder.

$$F = P \times A$$
$$= 5 \times 30$$
$$= 150 N$$

7 (h) Boyle's law

Boyle's Law states that the pressure of a fixed mass of gas is inversely proportional to its volume at a constant pressure. For example, if the volume is halved, the pressure is doubled; and if the volume is doubled, the pressure is halved.

Explanation: The reason for this effect is that a gas is made up of loosely spaced molecules moving at random. If a gas is compressed in a container, these molecules are pushed together; thus, the gas occupies less volume. The molecules, having less space in which to move, hit the walls of the container more frequently and thus exert an increased pressure, as shown in the figure 7.7.

Figure 7.7: Changes in pressure and volume of a gas

Other examples of change in gas volume and pressure

- The bubbles exhaled by a scuba diver grow as the approach the surface of the ocean. The pressure exerted by the weight of the water decreases with depth, so the volume of the bubbles increases as they rise.
- Deep sea fish die when brought to the surface. The pressure decreases as the fish are brought to the surface, so the volume of gases in their bodies increases, and pops bladders, cells, and membranes.

7 (i) Worked example

A sealed plastic bottle of containing 500 cm³ of air at atmospheric pressure of 101 kPa gets compressed as an aeroplane takes off and the cabin pressure increases. The bottle is compressed to 400 cm³. Calculate the pressure in the cabin as the aeroplane is taking off, assuming that the bottle temperature remains constant.

Applying Boyle's Law,
$P_1 \times V_1 = P_2 \times V_2$, at constant pressure
$500 \times 101 = P_2 \times 400$
$P_2 = (500 \times 101) / 400$
$P_2 = 126.25 kPa$

Chapter Eight
Energy Sources and Transfer of Energy

8 (a, b, e) Energy is the ability to do work. It is measured in Joules or Kilojoules. Energy can exist in different forms. It is a scalar quantity. Some forms of energy are discussed below.

Type of energy		Explanation
POTENTIAL ENERGY		Potential energy is stored energy and the energy in a body due to its position (gravitational).
CHEMICAL ENERGY		Chemical energy is the energy stored in the bonds of atoms and molecules. Biomass, petroleum, natural gas, propane and coal are examples of stored chemical energy.
NUCLEAR ENERGY		Nuclear energy is the energy stored in the nucleus of an atom. It is the energy that holds the nucleus together. Nuclear energy is utilized for power generation by fission reactions.
STORED MECHANICAL ENERGY		Stored mechanical energy is energy stored in objects by the application of a force. Compressed springs and stretched rubber bands are examples of stored mechanical energy.
GRAVITATIONAL POTENTIAL ENERGY		Gravitational potential energy is the energy due to its place or position. Water in a reservoir behind a hydropower dam is an example of gravitational potential energy. When the water is released to spin the turbines, it becomes kinetic energy.
KINETIC ENERGY		Kinetic energy is the energy in a body due to its state of motion. It is the motion of waves, electrons, atoms, molecules and substances.

RADIANT ENERGY		Radiant energy is electromagnetic energy that travels in transverse waves. Radiant energy includes visible light, x-rays, gamma rays and radio waves. Solar energy is an example of radiant energy.
THERMAL ENERGY		Thermal energy (or heat) is the internal energy In substances; it is the vibration and movement of atoms and molecules within substances. Geothermal energy is an example of thermal energy.
SOUND ENERGY		Sound is the movement of energy through substances in longitudinal (compression and rarefaction) waves.
ELECTRICAL ENERGY		Electrical energy Is the movement of electrons. Lightning and electricity are examples of electrical energy.

The table below shows the energy conversion in a few devices.

Device	Form of energy input	Intermediate forms of energy	Form of energy output
Petrol engine	Chemical	Sound and thermal	Kinetic
Electric motor	Electrical	Sound and thermal	Kinetic
Boiler and turbine	Thermal	Sound	Kinetic
Hydraulic pump	Mechanical	Kinetic	Potential
Generator	Mechanical	Sound and thermal	Electrical

> **Remember:**
> The principle of conservation of energy states that **energy cannot be created nor destroyed**. It can only be transformed from one form to another.

8 (c) Kinetic energy

The kinetic energy of a body is the energy due to its state of motion.

Kinetic Energy (K.E.) $= \frac{1}{2} Mass\ (m)\ x\ Velocity\ (V)^2$

Potential energy

The gravitational potential energy is the energy in a body due to its position.

Gravitational potential energy = mass x acceleration due to gravity x height of body from the ground

$$GPE = m\ x\ g\ x\ h$$

8 (d) Renewable and non-renewable energy resources

Renewable energy sources quickly replenish themselves and can be used again and again. For this reason they are sometimes called **infinite energy resources**.	**Non-renewable energy** sources cannot replenish themselves and cannot be used again and again. For this reason they are sometimes called **finite energy resources**.

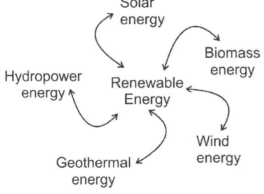

Figure 8.1: Renewable energy sources

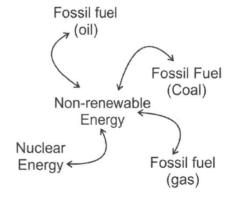

Figure 8.2: Non-renewable energy sources

Advantages of renewable energy	Advantages of non-renewable energy
• Potentially infinite energy supply. • Energy from biomass is a cheap and readily available source of energy. If biomass is replaced, it can be a long-term, sustainable energy source.	• Relatively cheap and provides large amounts of energy. • Relatively easy to transport, store and obtain.
Disadvantages of renewable energy	**Disadvantages of non-renewable energy**
• Installation can be costly and often damages ecosystems. • When burned, wood gives off atmospheric pollutants, including greenhouse gases. If trees are not **replanted** then wood is a non-renewable resource.	• Combustion of fossil fuel releases gases that cause acid rain and global warming. • Nuclear reactors are expensive to run. Nuclear waste is highly toxic, and needs to be safely stored for hundreds or thousands of years (storage is extremely expensive).

8 (f) Nuclear fusion and fission

Nuclear reactions involve release of binding energy when there is change in mass in the nucleus of an atom. The splitting of a large nucleus into two smaller nuclei is called nuclear fission and the formation of a larger nucleus by combining two smaller nuclei is called fusion. Both nuclear fission and fusion release large amounts of energy, which can be used to generate electricity.

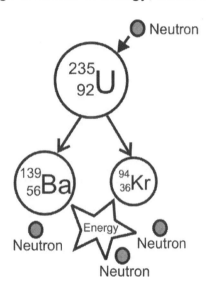

Figure 8.3: Nuclear fission of uranium nucleus   *Figure 8.4: Nuclear fission of hydrogen isotopes*

Advantages of generating electricity from nuclear energy
- A small amount of nuclear fuel can provide a large quantity of energy.
- Using nuclear energy reduces the dependence on fossil fuels.

Disadvantages of generating electricity from nuclear energy
- Nuclear reactors are expensive to maintain and install.
- Nuclear waste is highly toxic, and needs to be safely stored for hundreds or thousands of years (storage is extremely expensive).
- It is a non-renewable source of energy.

8 (g) Electricity generation

The block diagram in figure 8.5 shows the process of generating electricity from fossil fuel. The energy conversion taking place at each stage has been described.

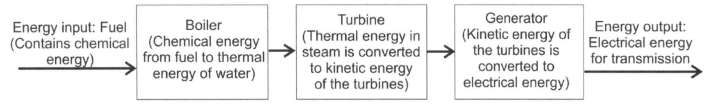

Figure 8.5: Block diagram of electricity generation from fossil fuel

8 (h) Environmental issues associated with power generation
- Burning of fossil fuels releases Sulfur dioxide and nitrogen oxides, which contribute to acid rain.
- Carbon dioxide released from burning of fossil fuels contributes to global warming.
- Wind farms used for generating electricity can cause ecological damage and spoil the landscape.

8 (i) Work

When a body is moved under the influence of a force, work is said to be done. The work done is the product of the force and the displacement of the body. Work is measured in Joules. If a force of 1 Newton moves a body over a distance of 1 metre then we say that 1 joule of work has been done. Work is a scalar quantity.

Work = force (F) × displacement (s)

8 (j) Efficiency of energy conversion

During conversion of energy, all the energy input into a system cannot be converted into useful output energy. The efficiency is always less than 1 or less than 100%. This is because some amount of energy is lost to the surrounding or converted into other forms that cannot be used. Efficiency has no unit. It may be expressed as a percentage.

$$\text{Efficiency} = \frac{useful\ energy\ output}{total\ energy\ input} \quad \text{or} \quad \text{\% Efficiency} = \frac{useful\ energy\ output}{total\ energy\ input} \times 100$$

8 (k, l) Efficiency of energy conversions in generating electricity

Figure 8.6 shows the energy conversion during generation of electrical energy. The efficiency of the system in converting chemical energy in fuel to electrical energy is calculated as below.

Efficiency = useful energy output / total energy input
$$= 160 / 600$$
$$= 0.26 \text{ or just } 26\%$$

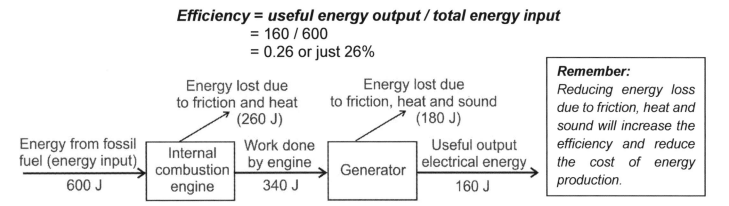

Remember:
Reducing energy loss due to friction, heat and sound will increase the efficiency and reduce the cost of energy production.

Figure 8.6: Efficiency of conversion in electricity generation

8 (m) Power

Power is the rate of doing work. It is measured in Watts (W). 1 Watt is the power when 1 Joule of work is done in 1 second.

$$Power = \frac{Work\ done}{Tme\ taken}$$

Worked example

A crane lifts a load of 900 N through a vertical height of 2.0 m in 12 s. The input power to the crane is 600 J / s.

Calculate the work done (energy output) by the crane.	Calculate the efficiency of the crane.
Work = force x displacement	*Efficiency = (energy output /energy input) x 100*
= 900 x 2 = 1800 J	*= (1800 / 7200) x 100*
	= 25%
Calculate the energy input.	
Energy = power x time	
= 600 x 12 = 7200 J	

Chapter Nine
Transfer of Thermal Energy

Cambridge 5054 syllabus specification 9 (a) *describe how to distinguish between good and bad conductors of heat.*
Cambridge 5054 syllabus specification 9 (b) *describe, in terms of the movement of molecules or free electrons, how heat transfer occurs in solids.*
Cambridge 5054 syllabus specification 9 (c) *describe convection in fluids in terms of density changes.*
Cambridge 5054 syllabus specification 9 (d) *describe the process of heat transfer by radiation.*
Cambridge 5054 syllabus specification 9 (e) *describe how to distinguish between good and bad emitters and good and bad absorbers of infra-red radiation.*
Cambridge 5054 syllabus specification 9 (f) *describe how heat is transferred to or from buildings and to or from a room.*
Cambridge 5054 syllabus specification 9 (g) *state and explain the use of the important practical methods of thermal insulation for buildings.*

9 (a) Heat

Heat is a form of energy. It is measured in Joules. Heat always flows from a hot body to a cold body. Conductors are bodies that transmit heat energy very rapidly. Insulators prevent the flow of heat.

Good conductors of heat will allow heat energy to travel rapidly through the body, as shown in figure 9.1.

Bad conductors of heat will slow down the transfer of heat through the body, as shown in figure 9.2.

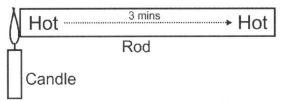

Figure 9.1: Good conductor of heat

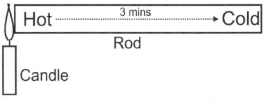

Figure 9.2: Bad conductor of heat

9 (b) Heat transfer in solids by conduction

The transfer of heat in solids occurs by conduction. Heat travels from a region of higher temperature (hot region) to a region of lower temperature(cold region) through the solid.

At the hot end, the molecules and free electrons vibrate faster (more kinetic energy). These high energy molecules or free electrons collide with the less energetic neighboring molecules and transfer heat energy to them. In this way heat energy is transferred from the hot end of the rod to the cold end of the rod by conduction, as illustrated in figure 9.3 and 9.4.

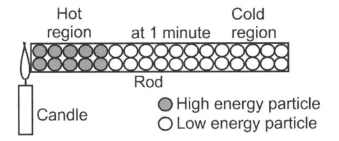

Figure 9.3: Particles near the heat source gain kinetic energy

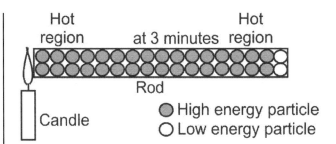

Figure 9.4: Particles transfer kinetic energy away from the source

9 (c) Transfer of heat in fluids by convection

Liquids and gases are fluids, as their particles can move from one place to another. When a fluid is heated, its molecules gain kinetic energy and move further apart. The volume increases and density decreases. This causes the less dense (high energy) molecules to move upward and the more dense (low energy) molecules to move downwards. So, heat is transferred from a hot region to a cooler region by the movement of molecules. This is termed as a convection current.

A convection current can be made visible by placing a crystal of purple potassium permanganate into a beaker containing water and heating the water with a candle or Bunsen burner, as shown in figure 9.5. the arrows show the movement of the purple dye through the water. This circulation of molecules is used to transfer heat from warmer to cooler regions by convection in liquids and gases.

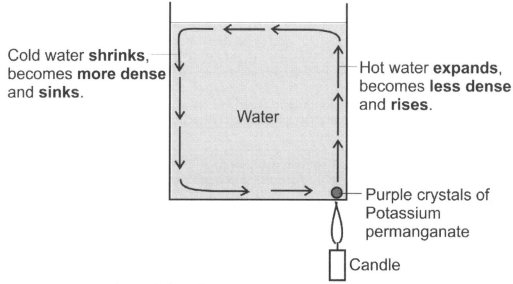

Cold water **shrinks**, becomes **more dense** and **sinks**.

Water

Hot water **expands**, becomes **less dense** and **rises**.

Purple crystals of Potassium permanganate

Candle

Figure 9.5: Making a convection current visible

9 (d) Heat transfer by radiation

All objects give out and take in **thermal radiation**, which is also called **infrared radiation**. The hotter an object is, the more infrared radiation it emits.

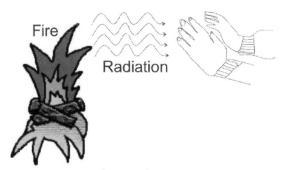

Fire

Radiation

Figure 9.6: Thermal energy by radiation

Infrared radiation is a type of electromagnetic radiation that involves waves. No particles are involved, unlike in the processes of conduction and convection, so radiation can even work through the vacuum of space. This is why we can still feel the heat of the Sun, although it is 150 million km away from the Earth.

9 (e) Absorbing and emitting infra-red radiation

Some surfaces are better than others at reflecting and absorbing infrared radiation, as shown in figure 9.7. The initial temperature of the container with the dull black surface and the shiny silvery surface is the same. The temperature increase in the black body is higher than in the shiny body when the heater is switched on. Similarly the black body cools more rapidly when the heater is switched off.

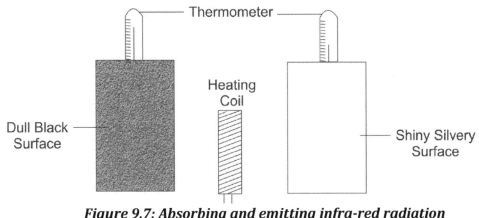

Figure 9.7: Absorbing and emitting infra-red radiation

> **Remember:**
> Black bodies with rough surfaces are good absorbers and good emitters of infra-red radiation. So, they heat up rapidly and cool down rapidly.
> White bodies with shiny surfaces are bad absorbers and bad emitters of infra-red radiation. So, they heat up slowly and cool down slowly

9 (f, g) Heat transfer to or from buildings and methods of thermal insulation

Most heat from a building is lost through the windows and roof by convection and conduction. Some heat is also lost through the floor by conduction. Heat energy also leaves the house by radiation through the walls, roof and windows.

Figure 9.8 summarizes some simple methods used to reduce heat loss from buildings and keep the buildings warm during extremely cold winters. This helps to reduce the cost of heating houses by using electrical energy.

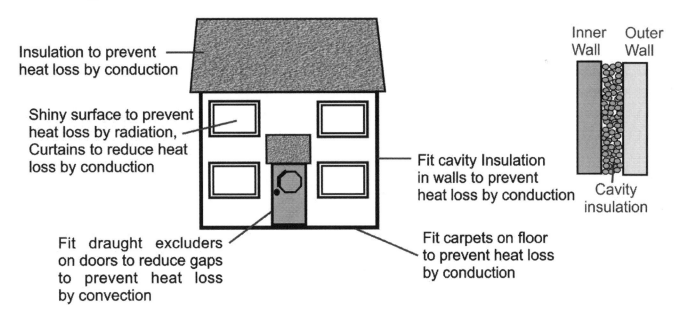

Figure 9.8: Reducing heat loss from buildings

Chapter Ten
Temperature

Cambridge 5054 syllabus specification 10 (a) explain how a physical property which varies with temperature may be used for the measurement of temperature and state examples of such properties.
Cambridge 5054 syllabus specification 10 (b) explain the need for fixed points and state what is meant by the ice point and steam point.
Cambridge 5054 syllabus specification 10 (c) discuss sensitivity, range and linearity of thermometers.
Cambridge 5054 syllabus specification 10 (d) describe the structure and action of liquid-in-glass thermometers (including clinical) and of a thermocouple thermometer, showing an appreciation of its use for measuring high temperatures and those which vary rapidly.

10 (a) Temperature

Temperature is the degree of hotness or coldness of a body. The atoms and molecules in a substance are in constant motion. Temperature is a measure of the average speed with which they move. More exactly it is a measure of their average kinetic energy. The higher the temperature, the faster the molecules move.

The physical properties, like density, length, electrical resistance, gas pressure vary with temperature. So, it is possible to measure temperature by measuring the values of these physical properties.

For example the density of mercury changes when its temperature is changed.

So in a confined tube, the height of the mercury can be used to measure the temperature, as shown in figure 10.1.

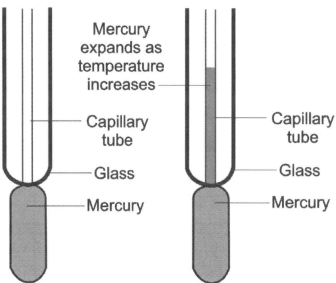

Figure 10.1: Thermal expansion of mercury (Thermometric property)

Temperature can be measured by using any physical properties that vary with temperature. Such properties are called thermometric properties. Examples of thermometric properties are

Thermometric property	Type of thermometer
Volume of a liquid	Liquid-in-glass thermometer
Length of solid	Bimetallic strip
Gas pressure	Constant volume gas thermometer
Electrical resistance	Resistance of a platinum resistor
Electromotive force	e.m.f of a thermocouple when the junctions are placed in different temperature

10 (b) Calibration of a thermometer

The mercury in glass thermometer needs to be calibrated to mark the scale on the glass tube. Figure 10.2 shows how the upper and lower fixed points of the thermometer are determined.

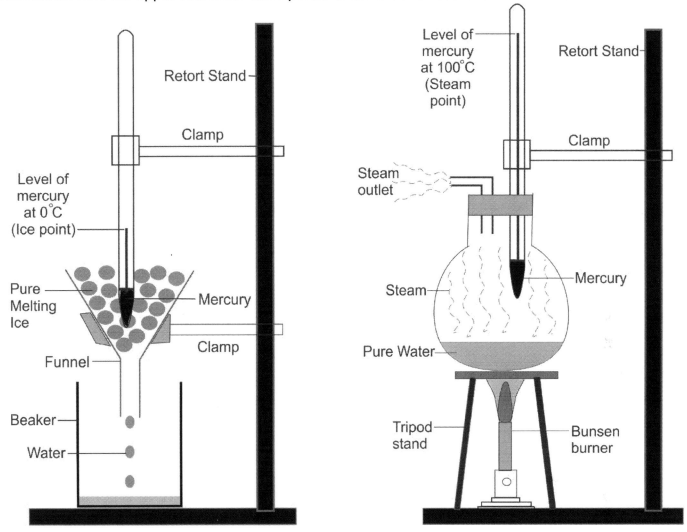

Figure 10.2: Fixing the Upper and lower fixed point of the liquid in glass thermometer.

Put thermometer into pure melting ice. After few minutes, when the mercury does not change its volume anymore, mark the position of the mercury in the tube. This is the ice point at 0^0 C.

Put the thermometer into steam. After few minutes, when the mercury does not change its volume anymore, mark the position of the mercury in the tube. This is the steam point at 100^0 C.

Divide the length between the ice point and steam point into 100 equal divisions, as shown in figure 10.3.

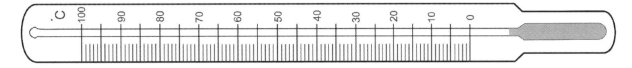

Figure 10.3: A calibrated liquid-in-glass thermometer

10 (c) Sensitivity, range and linearity of thermometers

Sensitivity refers to the ability to give a large response to a small change in temperature. A sensitive thermometer is able to detect small changes in temperature.

A sensitive thermometer will produce a large change in the length of the mercury column for every 1^0 **C change in temperature.** Typically, a thermometer with a thicker bore will be less sensitive since the change in length of the mercury column is lesser, as shown in figure 10.4.

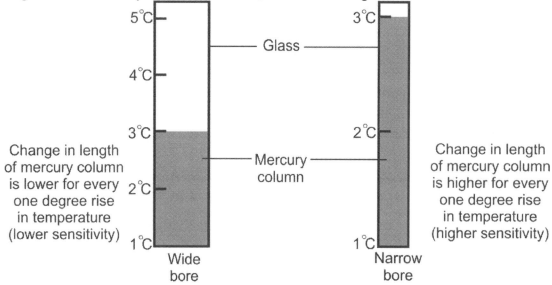

Figure 10.4: The diameter of the bore in the capillary tube affects sensitivity

Also, using a **large bulb in the thermometer can increase the sensitivity**. A large bulb will cause a big change in volume of the mercury, which will appear as a larger change in the length of mercury up the capillary tube.

Lastly, a liquid-in-glass thermometer may increase its sensitivity by choosing a liquid that expand more. Alcohol expands more than mercury, and would make a thermometer more sensitive than a mercury-in-glass thermometer.

Range

Range refers to the operating temperature at which the thermometer can be used.

A laboratory thermometer can measure from -10 °C to 110 °C and a clinical thermometer has a range between 30 ºC to 42 ºC, as shown in figure 10.5. The expanding liquid column may break the thermometer if the expansion exceeds beyond the maximum temperature. The range of a thermometer is affected by the length of the stem. The longer the length of the stem, the larger the range of the thermometer.

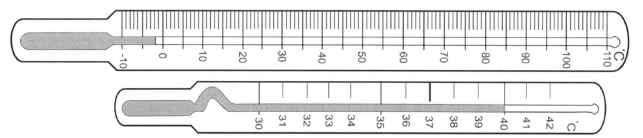

Figure 10.5: A laboratory and clinical thermometer with different range

A thermocouple which uses two metal junctions can have a large range from -200 °C to over 1000 °C. This make thermocouple a very versatile thermometer and is suitable in many situations.

Linearity (Uniformity)

The expansion of the liquid should be linear (uniform, the same) for all different temperatures measured.

10 (d) Clinical thermometer

A liquid-in-glass thermometer uses the principle of thermal expansion of a liquid. The liquid is held in a bulb and the liquid expands and rises into a capillary tube in a vacuum. The higher the temperature, the higher is the rise in the level of the liquid.

The thermometric liquid may be alcohol or mercury. The advantages and disadvantages of each liquid is listed below.

Mercury as a thermometric liquid

Advantages	Disadvantages
Mercury is a good conductor of heat, so the whole liquid reaches the temperature of the surrounding easily. It also expands rapidly.	Mercury is poisonous.
It does not wet (cling to) the sides of the glass tube.	Its expansivity is fairly low.
It expands uniformly and improves the linearity.	Mercury is expensive.
It is opaque and clearly visible.	It has a high freezing point (-39 °C) and cannot be used to record very low temperatures.

Alcohol as a thermometric liquid

Advantages	Disadvantages
It expands uniformly.	It wets the glass tube (sticks to glass).
It has a low freezing point (-115 °C) and can be used to measure low temperatures.	It has a low boiling point (78 °C) and cannot record higher temperatures.
It has a large expansivity.	It does not react quickly to changes in temperature.
It is cheap and is a safe liquid.	It cannot be seen easily and needs to be coloured.

Figure 10.6 shows a clinical thermometer and some of the features that enable it to perform effectively.

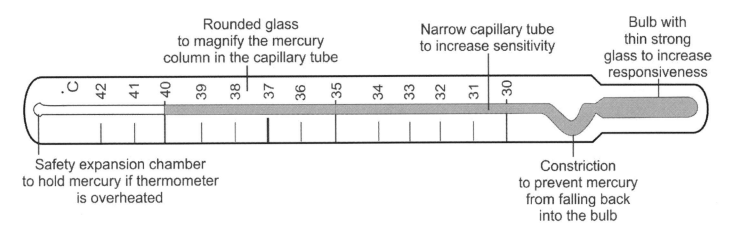

Figure 10.6: A clinical thermometer

Thermocouple

The thermocouple consists of two wires made of different metals such as copper or iron.

The ends of the wires are joined together to form two junctions. If the two junctions are at different temperatures, that is, one junction is hot and other is cold, a small e.m.f. (a voltage) is produced, the greater the difference in temperature, the greater is the e.m.f. produced across the ends of the two junctions, as shown in figure 10.7.

If one junction is kept at a fixed cold temperature (0 ^{0}C), then the other junction can be used as a tiny probe to measure temperatures other than 0 ^{0}C.

No voltage is produced when the two junctions are placed at the same temperature.

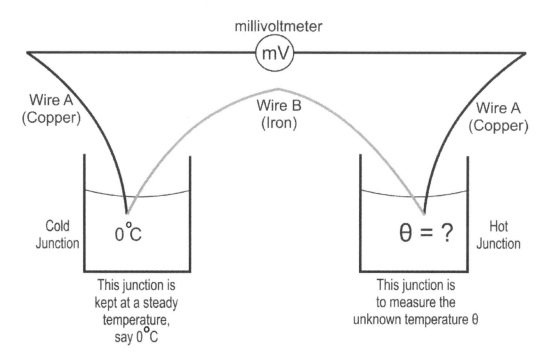

Figure 10.7: A thermocouple

Advantages of a thermocouple are

- since it is electrical, it can be read from a remote site. eg, inside a furnace.
- It can measure a very large temperature range of -200 ^{0}C to 1500 ^{0}C by choosing suitable types of metals for wires A and B.
- a thermocouple can record rapid changes in temperature and can measure localized temperature (as the junctions are very small, it can measure the temperature exactly at a particular location or point on a surface).

Chapter Eleven
Thermal Properties of Matter

Cambridge 5054 syllabus specification 11 (a) describe a rise in temperature of a body in terms of an increase in its internal energy (random thermal energy).
Cambridge 5054 syllabus specification 11 (b) define the terms heat capacity and specific heat capacity.
Cambridge 5054 syllabus specification 11 (c) recall and use the formula thermal energy = mass × specific heat capacity × change in temperature.
Cambridge 5054 syllabus specification 11 (d) describe melting/solidification and boiling/condensation in terms of energy transfer without a change in temperature.
Cambridge 5054 syllabus specification 11 (e) state the meaning of melting point and boiling point.
Cambridge 5054 syllabus specification 11 (f) explain the difference between boiling and evaporation.
Cambridge 5054 syllabus specification 11 (g) define the terms latent heat and specific latent heat.
Cambridge 5054 syllabus specification 11 (h) explain latent heat in terms of molecular behaviour.
Cambridge 5054 syllabus specification 11 (i) calculate heat transferred in a change of state using the formula thermal energy = mass × specific latent heat.
Cambridge 5054 syllabus specification 11 (j) describe qualitatively the thermal expansion of solids, liquids and gases.
Cambridge 5054 syllabus specification 11 (k) describe the relative order of magnitude of the expansion of solids, liquids and gases.
Cambridge 5054 syllabus specification 11 (l) list and explain some of the everyday applications and consequences of thermal expansion.
Cambridge 5054 syllabus specification 11 (m) describe qualitatively the effect of a change of temperature on the volume of a gas at constant pressure.

11 (a) Thermal energy
According to the Kinetic theory of matter, the average kinetic energy (K.E) of molecules is proportional to the absolute temperature (T).

Average Kinetic Energy of particles α Absolute Temperature

K.E α T

Matter becomes hotter when heat is transferred to them. This energy is absorbed by the molecules which result in the molecules gaining more kinetic energy, hence moving faster. The faster moving molecules cause the temperature of the system to rise.

11 (b) Heat capacity
Heat capacity is the energy needed to raise the temperature of a body by 1^0C (or 1 K).

For example, 2 kg of water will require 8400 J of energy to warm up from 25 ^{0}C to 26^0C and 3 kg of water will need 12600 J of heat energy to heat up from 25 ^{0}C to 26^0C.

Likewise, 0.5 kg of aluminum will require 455 J of heat to raise its temperature by 1^0C.

11 (c) Specific Heat Capacity
Specific Heat capacity is the energy needed to raise the temperature of a 1 kg body by 1^0C (or 1 K).
The SI unit of specific heat capacity is J/kg K.

For example, the specific heat capacity of water is 4200 Jkg⁻¹ ⁰C⁻¹. This means that 4200 J of heat energy will have to be provided to increase the temperature of 1kg of water by 1°C, as shown in figure 11.1.

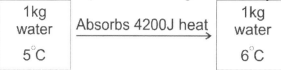

Likewise, the specific heat capacity of copper is 390 Jkg⁻¹K⁻¹. This means that 390 J of heat energy will have to be provided to increase the temperature of 1kg of copper by 1⁰C.

This relationship between the energy supplied, mass and the temperature change can be expressed in the formula:

$$E = mc\Delta\theta$$

Where,
E = thermal energy supplied
m = mass of metal block
c = specific heat capacity
$\Delta\theta$ = increase in temperature of the block ($\theta_2 - \theta_1$)
θ_2 = Final temperature
θ_1 = Initial temperature

Determining the specific heat capacity of a solid

An electric heating coil connected to a D.C supply of 12V and carrying a current of 5A for 60 seconds is used to heat a metal block of mass 0.70 kg. the temperature of the block increases from 25 ⁰C to 37 ⁰C. assuming that all the energy from the coil is absorbed by the metal block, calculate the specific heat capacity of the metal. The arrangement is shown in figure 11.1.

$E = mc\Delta\theta$

$m = 0.70 kg$

$c =$

$\Delta\theta = 12 \ ^\circ C$

$V = 12 V$

$I = 5 A$

$t = 60 \ s$

So, $c = (VIt) / m \ \Delta\theta$

$= (12 \times 5 \times 60) / (0.70 \times x12)$

$= 3600 / 8.4$

$= 429 \ J \ Kg^{-1} \ K^{-1}$

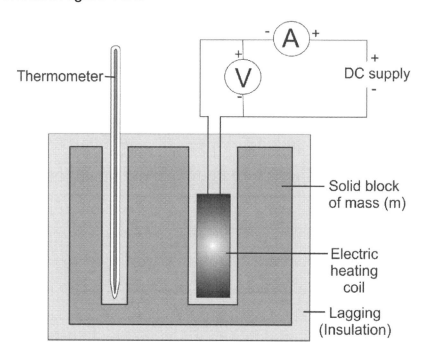

Figure 11.1: A lagged calorimeter

11 (d) (e) Melting

The solid changes to its liquid form at a constant temperature, called the melting point. During this period, the thermal energy is used to do work to break the intermolecular bonds between the molecules of the solid. There is no change in kinetic energy during this period, so temperature remains constant.

Melting point

The constant temperature at which a solid changes to its liquid form. Every substance has a fixed melting point and the melting point can be used to identify the substance or to check its purity.

Boiling

The liquid changes to its gaseous form at a constant temperature, called the boiling point. During this period, the thermal energy is used to do work separate the molecules as well as in pushing back the surrounding atmosphere. There is no change in kinetic energy during this period, so temperature remains constant.

Boiling point

The constant temperature at which a liquid changes to its gaseous form. Every substance has a fixed boiling point and the boiling point can be used to identify the substance or to check its purity.

Heating curve

The heating curve shows the changes in temperature of a solid over time, as shown in figure 11.2.

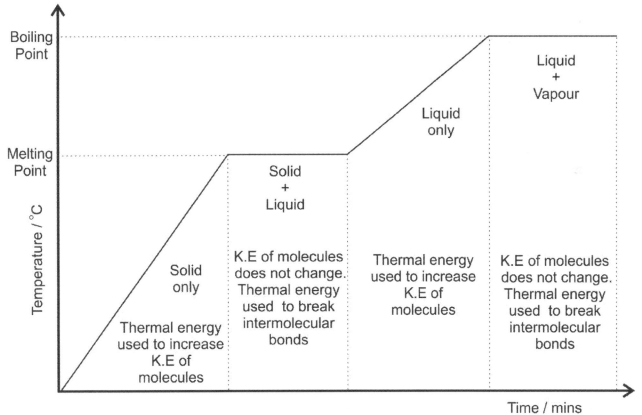

Figure 11.2: Heating curve

11 (f) Differences between boiling and evaporation.

Boiling	Evaporation
Change of state from liquid to vapour at a fixed temperature	Change of state from liquid to vapour at all temperatures
Occurs from all parts of the liquid, so bubbles are formed.	Occurs only from the surface of the liquid, so no bubbles form.
Quicker process	Slower process
Thermal energy must be supplied by an energy source	Thermal energy supplied by the surroundings.

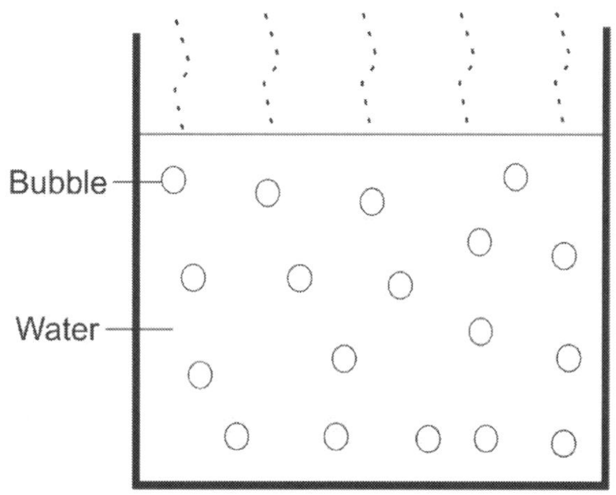

Figure 11.3: Boiling

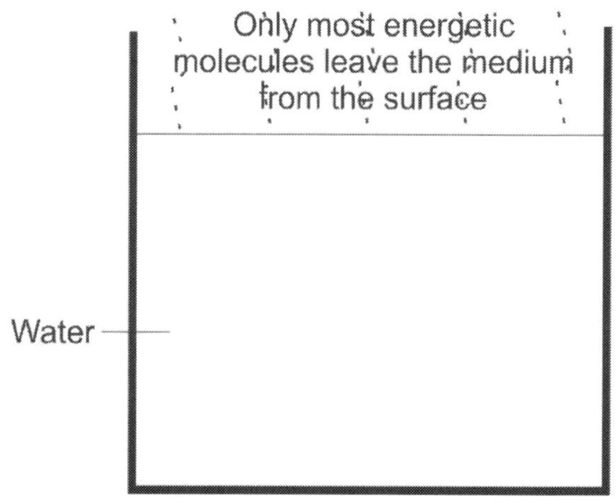

Only most energetic molecules leave the medium from the surface

Figure 11.4: Evaporation

11 (g, i) Latent heat
The heat energy needed to change the state of a substance at a constant temperature.

Specific Latent heat
The heat energy needed to change the state of a 1kg of a substance at a constant temperature.

Specific Latent heat of vapourisation
The heat energy needed to change the state of a 1kg of a substance from a liquid to vapour at a constant temperature.

Specific Latent heat of fusion
The heat energy needed to change the state of a 1kg of a substance from a solid to liquid at a constant temperature.

thermal energy = mass × specific latent heat

11 (h) Molecular behavior and latent heat.
Latent means hidden. The heat energy being supplied to the medium does not cause a change in temperature and appears to be having no effect on the medium. However, during melting or boiling, the heat energy is used to change the state of the medium. The energy is used to do work and to break the intermolecular forces between the molecules.

11 (j, k, l) Thermal expansion is solids, liquids and gases

When a substance is heated, the molecules gain kinetic energy and vibrate with a greater amplitude. This causes the particles to move further apart and the substance expands. This is called thermal expansion. Thermal expansion is illustrated in a solid in figure 11.5.

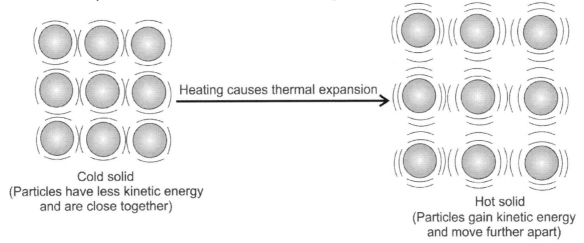

Figure 11.5: Thermal expansion in solids

When metals and other solids that have high thermal expansion are used to make bridges or railway lines, some gaps must be available for thermal expansion and contraction during day and night, as shown in figure 11.6.

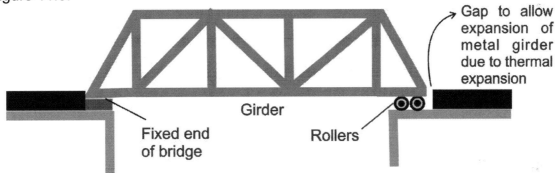

Figure 11.6: Thermal expansion and contraction in bridges

Liquids also expand on with an increase in temperature, as shown in figure 11.7.

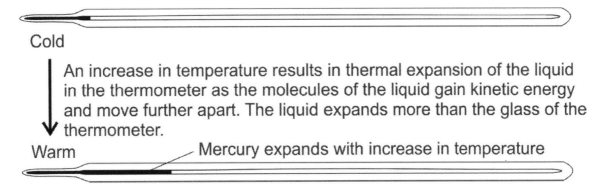

Figure 11.7: Thermal expansion of liquid mercury

Gases also expand as temperature increases, as the molecules move further apart, as illustrated when the air in a balloon is heated. The air expands and becomes less dense, causing the balloon to rise.

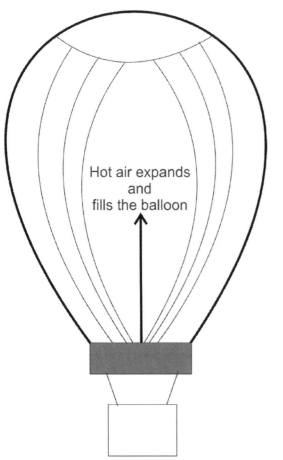

Figure 11.8: Thermal expansion of gases

> **Remember:**
> *Expansion in solids is minimum and expansion in gases is maximum.*
>
> *Solids expansion < liquid expansion < gas expansion*

> **11 (m) Remember:**
> *Charle's Law states that the increase in volume of a fixed mass of gas is proportionate to its absolute temperature, at a constant pressure.*
>
> *Volume α Absolute Temperature, at constant Pressure*
>
> $$\frac{V1}{T1} = \frac{V2}{T2}$$, at constant pressure

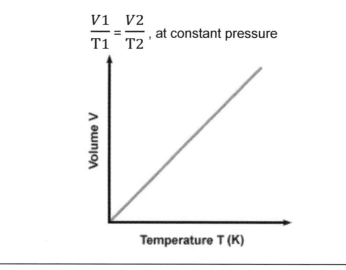

Chapter Twelve
Kinetic Model of Matter

Cambridge 5054 syllabus specification 12 (a) *state the distinguishing properties of solids, liquids and gases.*
Cambridge 5054 syllabus specification 12 (b) *describe qualitatively the molecular structure of solids, liquids and gases, relating their properties to the forces and distances between molecules and to the motion of the molecules.*
Cambridge 5054 syllabus specification 12 (c) *describe the relationship between the motion of molecules and temperature.*
Cambridge 5054 syllabus specification 12 (d) *explain the pressure of a gas in terms of the motion of its molecules.*
Cambridge 5054 syllabus specification 12 (e) *describe evaporation in terms of the escape of more energetic molecules from the surface of a liquid.*
Cambridge 5054 syllabus specification 12 (f) *describe how temperature, surface area and draught over a surface influence evaporation.*
Cambridge 5054 syllabus specification 12 (g) *explain that evaporation causes cooling.*

12 (a, b) Properties of solids, liquids and gases

Property	Solid	Liquid	Gas
Shape	Fixed	Takes the shape of the container	Fills any space that is available
Volume	Definite volume (for a fixed mass of a substance at a specific temperature)	Definite volume (for a fixed mass of a substance at a specific temperature)	Occupies the entire space available
Density	Generally high	Generally lower than solids	Lower than solids and liquids
Intermolecular spaces	Intermolecular spaces are very small. Particles cannot change places and vibrate about their mean position. This makes solids less compressible.	Intermolecular spaces are larger than in solids. Particles can change places within the medium. So, liquids are slightly more compressible than solids.	Intermolecular spaces are very large. Particles move freely in all directions. Gases are highly compressible.
	Figure 12.1	*Figure 12.2*	*Figure 12.3*
Intermolecular forces	Intermolecular forces are very high and hold the particles together.	Intermolecular forces are less than in solids, so particles can move further apart and change places.	Intermolecular forces are very low and particles can move randomly in all directions and fill all available spaces.

12 (c) Motion of molecules and temperature.

According to the kinetic theory of matter, the average kinetic energy of particles is proportional to the absolute temperature.

Kinetic Energy α Average Temperature

So, as a substance is heated the particles gain kinetic energy and vibrate more rapidly and move further apart from each other. Temperature increases as the average kinetic energy of the particles increases. This results in thermal expansion of the substance. The thermal expansion for solids is minimum and the thermal expansion of gases is maximum.

12 (d) Pressure of a gas in terms of the motion of its molecules

The particles of a gas are constantly in motion and collide with the walls of the container. The force exerted by the particles per unit area of the container at any given moment is the pressure.

If the particles are at a higher temperature, the force exerted by the particles will be more as they have more kinetic energy and collide more frequently with the walls of the container, as explained in figure 12.4.

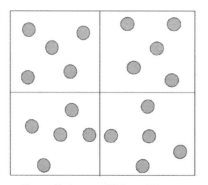

The gas is heated at a constant volume. The particles gain kinetic energy and bombard the walls of the container more frequently with a greater force.

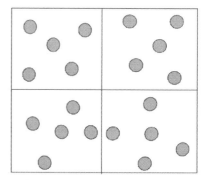

5 particles collide with one unit area of the container at any given moment.

5 particles collide with one unit area of the container with a greater force.

Figure 12.4: Changes in pressure of a gas with temperature

Likewise, if the gas in the container is compressed at a constant temperature, the area over which the particles are exerting the force decreases and the pressure will increase, as explained in figure 12.5.

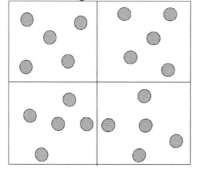

Volume of gas is halved and pressure is doubled

(Mass of gas is fixed and temperature is constant)

10 particles collide with one unit area of the container at any given moment. So, pressure is doubled.

5 particles collide with one unit area of the container at any given moment.

Figure 12.5: Changes in pressure of a gas with volume

12 (e) Evaporation

According to the kinetic theory of matter, the particles in a liquid are in continuous random motion. Some molecules have slightly more energy than others. The more energetic molecules escape from the surface of the liquid by breaking away from the intermolecular forces in the liquid. The less energetic molecules stay back in the medium and the liquid cools down as the average kinetic energy of the particles in the medium decreases.

12 (g) Evaporation causes cooling. When methylated spirit evaporates from the surface of the skin, the most energetic molecules evaporate and the less energetic molecules are left behind. As more molecules gain energy from the skin and evaporate, the skin cools down as the skin loses heat.

12 (f) Factors which increase the rate of evaporation are:

Surface area

The larger the surface of the liquid exposed to the atmosphere, the higher is the rate of evaporation. Figure 12.6 shows that water evaporates more rapidly from a wide mouthed basin than from a narrow mouthed bucket. As evaporation occurs only from the surface and a larger surface is exposed in the basin, more water molecules leave the medium at any given moment.

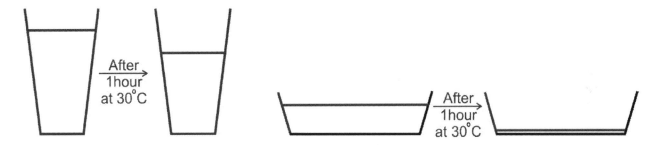

Figure 12.6: Rapid evaporation from a larger surface

Temperature

As temperature increases the rate of evaporation increases.

Draught or wind

An increase in wind velocity increases the rate of evaporation.

Example: Clothes dry quickly on a hot windy day, as shown in figure 12.7.

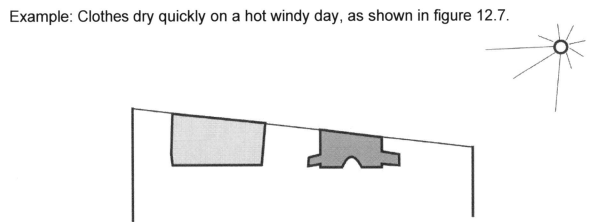

Figure 12.7: clothes dry quickly on a hot, windy day

Cambridge 5054 syllabus specification 13 (a) describe what is meant by wave motion as illustrated by vibrations in ropes and springs and by experiments using a ripple tank.
Cambridge 5054 syllabus specification 13 (b) state what is meant by the term wavefront.
Cambridge 5054 syllabus specification 13 (c) define the terms speed, frequency, wavelength and amplitude and recall and use the formula velocity = frequency × wavelength.
Cambridge 5054 syllabus specification 13 (d) describe transverse and longitudinal waves in such a way as to illustrate the differences between them.
Cambridge 5054 syllabus specification 13 (e) describe the use of a ripple tank to show
(1) reflection at a plane surface,
(2) refraction due to a change of speed at constant frequency.
(f) describe simple experiments to show the reflection and refraction of sound waves.

13 (a) A wave is a disturbance in a medium which causes the particles to vibrate about their mean positions and transfer energy to adjacent particles. The energy is carried away from the wave source.

Waves are of two main types:
- Longitudinal waves
- Transverse waves

13 (d) Longitudinal waves

In a longitudinal wave, the direction of vibration of particles is parallel to the direction of propagation of the wave, as shown in figure 13.1.

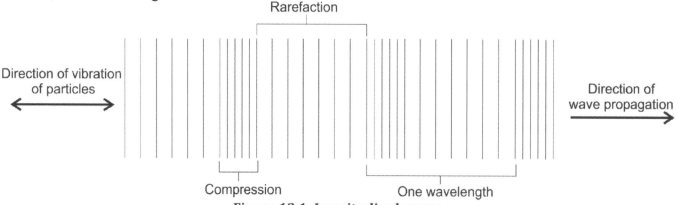

Figure 13.1: Longitudinal waves

The formation of longitudinal waves can easily be demonstrated by moving a slinky spring as shown in figure 13.2.

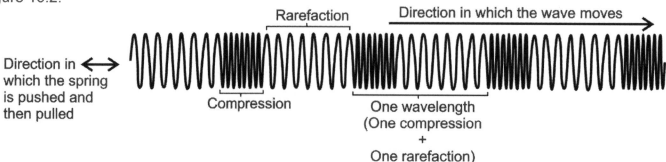

Figure 13.2: Longitudinal waves in a slinky spring

Transverse waves

In a transverse wave, the direction of vibration of particles is perpendicular to the direction of propagation of the wave, as shown by the rope in figure 13.3.

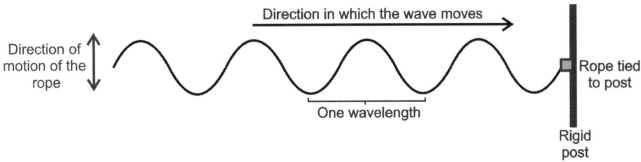

Figure 13.3: Transverse waves in a rope

A transverse wave is shown in figure 13.4.

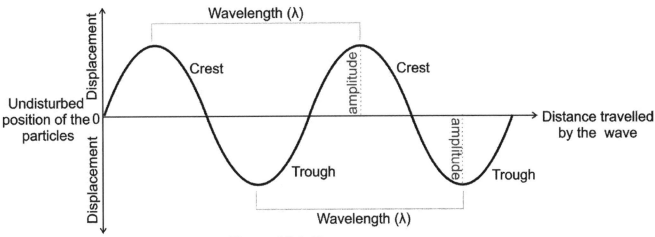

Figure 13.4: Transverse waves

13 (b) Wavefront

The particles in a medium vibrate about their mean positions as shown in figure 13.5. The particles which are in the same phase have the same direction and speed. For example particles B and F are in the same phase. Likewise, particles A, E and I are all in the same phase.

A wavefront is an imaginary line joining all the points within the wave that are in the same phase. It usually represents the top or the crest or the bottom of the trough, as illustrated in figure 13.6 and 13.7.

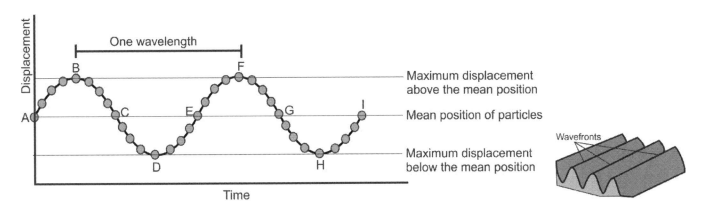

Figure 13.5: Oscillation of particles in a transverse wave and wavefronts

The top of every wave in a ripple tank can be joined by lines called as wavefronts. Using a paddle in the ripple tank will produce plane transverse waves and using a spherical metal bob to generate waves in the ripple tank produces circular transverse waves.

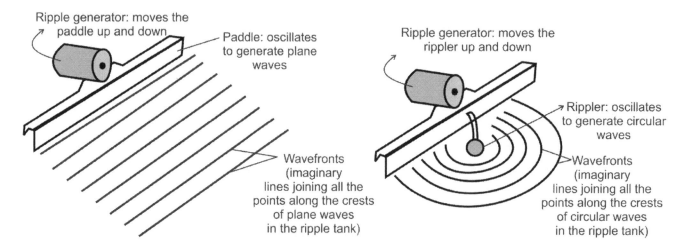

Figure 13.6: Wavefronts for plane waves   Figure 13.7: Wavefronts for circular waves

13 (c) Amplitude(a): The maximum displacement of the particles in the medium from their mean position of rest. The SI unit is metre (m). The amplitude is illustrated in figure 13.8 and 13.9.

Wavelength(λ): The shortest distance between two consecutive crests or troughs. The SI unit is metre (m). the wavelength is illustrated in figure 13.8.

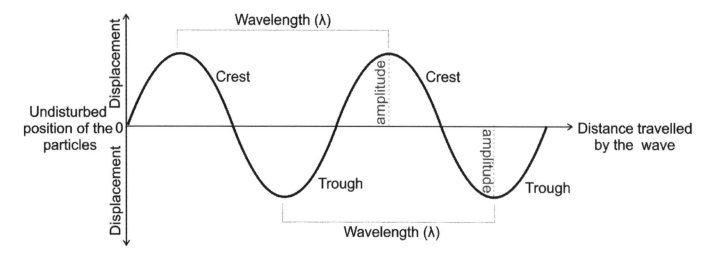

Figure 13.8: Graph showing displacement of oscillating particles and distance travelled by the wave

Frequency (f): The number of complete waves produced in one second. The SI unit is Hz (Hertz).

Wave speed (v): The distance travelled by the wave in one second. The SI unit is m/s.

Time period (T): the time taken to produce one complete wave. The SI unit is seconds (s). The time period is illustrated in figure 13.9.

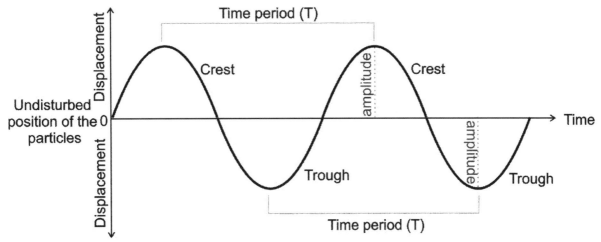

Figure 13.9: Graph showing displacement of oscillating particles and time

13 (c) Worked example

The velocity of a wave is determined by the frequency and the wavelength of the wave. Calculate the velocity of a 5 Hz wave with a wavelength of 0.2m.

velocity* = *frequency* × *wavelength

$$V = f \times \lambda$$
$$= 5 \times 0.2$$
$$= 1 m/s$$

Ripple tank

A ripple tank is shown in figure 13.10. The vibrator vibrates the paddle or bob to create plane waves or circular waves respectively. The paddle or bob creates a disturbance in the liquid medium (water) and initiates wave formation in the tank. The transverse waves are set up in the water and transmitted away from the source of production.

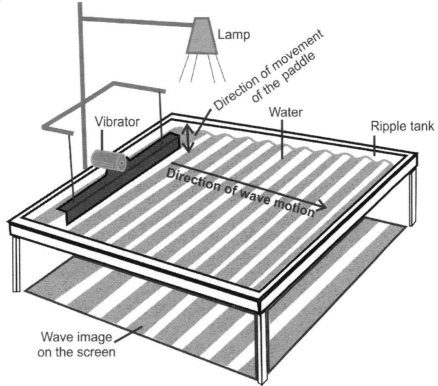

Figure 13.10: Ripple tank

13 (e) Reflection of waves at a plane surface

The transverse waves are represented by wavefronts. The distance between two consecutive wavefronts is called as the wavelength. As the waves move towards a glass slab in the ripple tank, as shown in figure 13.11, the waves get reflected from the surface of the glass slab.

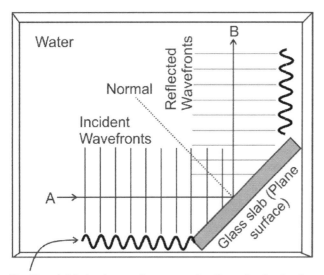

Remember:
The angle of incidence (angle between A and the normal) of the waves is equal to the angle of reflection (angle between B and the normal). The wavelength, amplitude, speed and frequency of the waves remain unchanged after reflection.

Profile (side) view of waves in the ripple tank. Notice that the wavefronts are imaginary lines along the top of each crest. The distance between two consecutive crests is the wavelength of the wave. So, the wavelength is also the distance between two consecutive wavefronts.

Figure 13.11: Reflection of waves in the ripple tank

Refraction due to a change of speed at constant frequency

As the wave moves towards the submerged glass slab, the depth decreases and the wave slows down and changes direction (refraction), as shown in figure 13.12. The wavelength of the refracted waves decreases and the amplitude increases. The frequency of the waves remains unchanged.

Top (Bird's eye) view of ripple tank

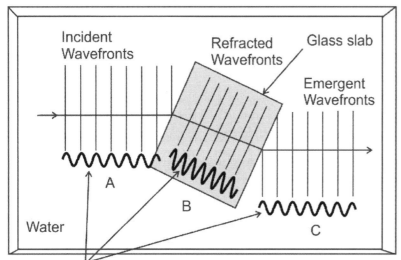

Side (profile) view of ripple tank

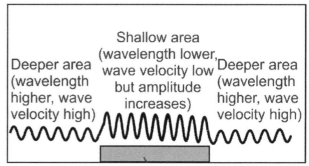

Profile (side) view of waves in the ripple tank. Notice that the wavefronts are imaginary lines along the top of each crest. The distance between two consecutive crests is the wavelength of the wave. So, the wavelength is also the distance between two consecutive wavefronts.

Figure 13.12: Refraction of waves in the ripple tank

13 (f) Experiment to show the reflection of sound waves

The equipment used to demonstrate reflection of sound waves is shown in figure 13.13. The mechanical stop watch produces a ticking sound. The sound waves travel towards the wooden board through a cardboard tube. The sound waves are reflected by wooden board. A soft wood board is placed between the stopwatch and the observer to prevent sound from the source travelling directly to the observer. The sound is the loudest when the angle of incidence is equal to the angle of reflection.

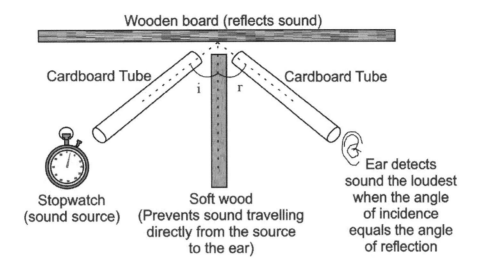

Figure 13.13: Reflection of sound waves

Experiment to show the refraction of sound waves

Sound waves travel through warm air faster than through cold air. During the night sound waves are refracted as sound waves travel at different speeds through warm and cold air as shown in figure 13.14. the sound waves from the source are refracted towards the observer and the sound of the train is loud and clear.

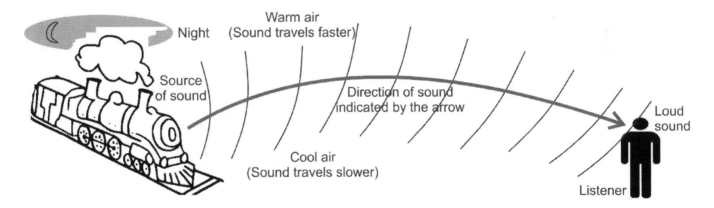

Figure 13.14: Refraction of sound waves at night

During the day sound waves are refracted as sound waves travel at different speeds through warm and cold air as shown in figure 13.15. The sound waves from the source are refracted away from the observer and the sound of the train is very faint. In both cases the observer is at the same distance away from the train.

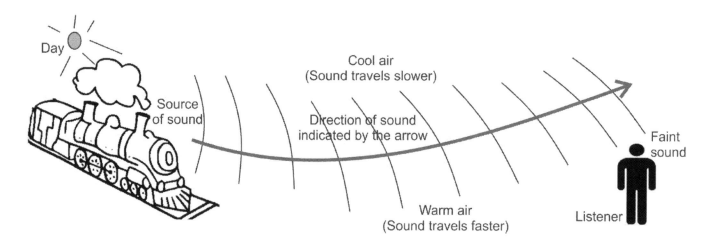

Figure 13.15: Refraction of sound waves in the day

Chapter Fourteen
Light

Cambridge 5054 syllabus specification 14 (a) *define the terms used in reflection including normal, angle of incidence and angle of reflection.*

Cambridge 5054 syllabus specification 14 (b) *describe an experiment to illustrate the law of reflection.*

Cambridge 5054 syllabus specification 14 (c) *describe an experiment to find the position and characteristics of an optical image formed by a plane mirror.*

Cambridge 5054 syllabus specification 14 (d) *state that for reflection, the angle of incidence is equal to the angle of reflection and use this in constructions, measurements and calculations.*

Cambridge 5054 syllabus specification 14 (e) *define the terms used in refraction including angle of incidence, angle of refraction and refractive index.*

Cambridge 5054 syllabus specification 14 (f) *describe experiments to show refraction of light through glass blocks.*

Cambridge 5054 syllabus specification 14 (g) *recall and use the equation sin i /sin r = n.*

Cambridge 5054 syllabus specification 14 (h) *define the terms critical angle and total internal reflection and recall and use the formula sin c = 1/n.*

Cambridge 5054 syllabus specification 14 (i) *describe experiments to show total internal reflection.*

Cambridge 5054 syllabus specification 14 (j) *describe the use of optical fibres in telecommunications and state the advantages of their use.*

Cambridge 5054 syllabus specification 14 (k) *describe the action of thin lenses (both converging and diverging) on a beam of light.*

Cambridge 5054 syllabus specification 14 (l) *define the term focal length.*

Cambridge 5054 syllabus specification 14 (m) *draw ray diagrams to illustrate the formation of real and virtual images of an object by a converging lens, and the formation of a virtual image by a diverging lens.*

Cambridge 5054 syllabus specification 14 (n) *define the term linear magnification and *draw scale diagrams to determine the focal length needed for particular values of magnification (converging lens only).*

Cambridge 5054 syllabus specification 14 (o) *describe the use of a single lens as a magnifying glass and in a camera, projector and photographic enlarger and draw ray diagrams to show how each forms an image.*

Cambridge 5054 syllabus specification 14 (p) *draw ray diagrams to show the formation of images in the normal eye, a short-sighted eye and a long-sighted eye.*

Cambridge 5054 syllabus specification 14 (q) *describe the correction of short-sight and long-sight.*

14 (a) Reflection

Light rays are represented by arrows to show the direction of light. Figure 14.1 shows the reflection of a single ray of light from a plane mirror.

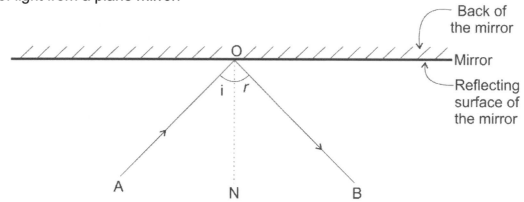

Figure 14.1: Reflection of light

The normal is a line (ON) which is perpendicular to the reflecting surface and drawn at the point of incidence (O).
Angle of incidence (AON): The angle between the incident ray (AO) and the normal (ON).
Angle of reflection (BON): The angle between the reflected ray (BO) and the normal (ON).

Laws of reflection
- The angle of incidence (i) is equal to the angle of reflection (r).

$$i = r$$

- The incident ray, the normal and the reflected ray all lie on the same plane at the point of incidence.

14 (b) Experiment to illustrate the laws of reflection

1. Draw a straight line XY on a white sheet of paper.
2. Place a plane mirror along the line XY.
3. Use a ray box which emits a narrow beam of red light (monochromatic light). Focus the beam of red light on to the mirror as shown in figure 14.2 and mark two points P_1 and P_2 along the beam with a pencil.
4. Observe the reflected beam and mark two points P_3 and P_4 along the reflected beam.
5. Remove the mirror and draw a line joining P_1 and P_2. Extend the line until it touched XY. This is the point of incidence (O).
6. Draw a normal (perpendicular) at the point of incidence.
7. Draw a line from the point of incidence passing through P_3 and P_4.
8. Measure the angle of incidence (i) and angle of reflection (r).

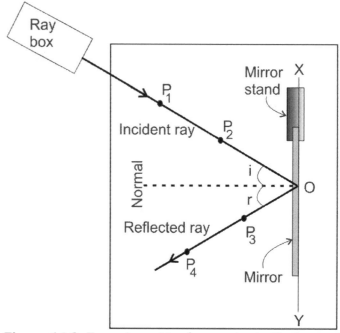

Figure 14.2: Experiment to demonstrate the laws of reflection

14 (c) (d) Position and characteristics of an optical image formed by a plane mirror

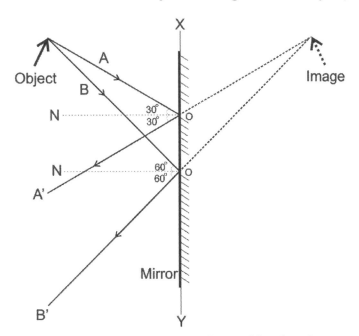

Figure 14.3: Features of an image formed by the plane mirror

The image formed by a plane mirror have the following features.

- The image is virtual (cannot be obtained on a screen).
- The distance between the object and the plane mirror is the same as the distance between the image and the plane mirror.
- The image is laterally inverted.
- The size of the image is the same as the size of the object. So, magnification is equal to 1.

14 (e) Refraction

Refraction is the bending of light as it passes from one medium to another.
Figure 14.4 shows the direction in which a light ray travels as it passes from one medium to another.

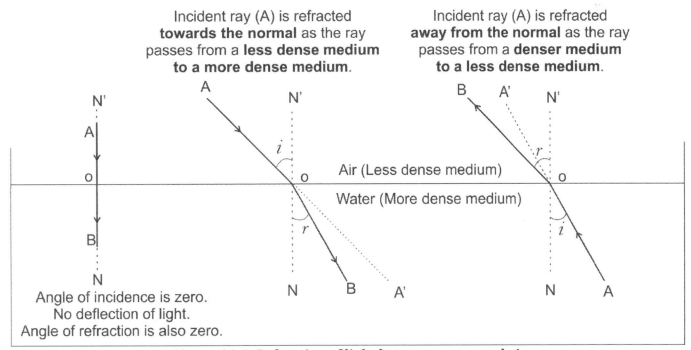

Figure 14.4: Refraction of light between water and air

14 (e) Refer to figure 14.5 to understand the following terms.

Angle of incidence (i): The angle between the incident ray (A) and the normal (N') is called as the angle of incidence.

Angle of refraction (r): The angle between the refracted ray (B) and the normal (N) is called as the angle of refraction.

Normal (NN'): The normal is a perpendicular line drawn at the point of incidence (O).

A' represents the path that would be followed by the ray if refraction did not occur.

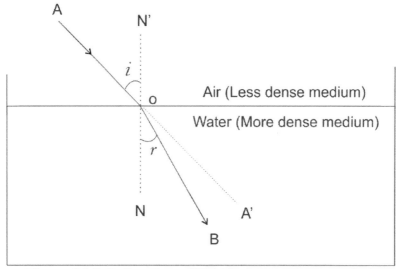

Figure 14.5: Refraction of light between water and air

14 (g) Refractive index
The refractive index is the ratio of *sin i* to *sin r*. Since it is a ratio, there is no unit.

$$\text{Refractive index (n)} = \frac{Sin\ i}{Sin\ r}$$

Where,
i is the angle of incidence, and
r is the angle of refraction

Worked example

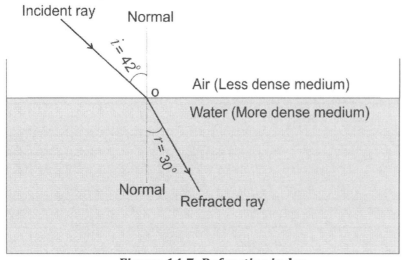

Figure 14.7: Refractive index

Figure 14.7 shows the angle of incidence (42°) and angle of refraction (30°) for a given ray of light passing from air into a glass block. Find the refractive index of water.

$$\text{Refractive index (n)} = \frac{Sin\ i}{Sin\ r}$$
$$= \sin 42\ /\ \sin 30$$
$$= 1.33$$

Note that refractive index has no units.

14 (f) Experiment to show refraction of light through a glass block

Procedure

1. Place a rectangular glass block on a white sheet of paper and draw the outline (ABCD) using a sharp pencil.

2. Use a ray box which emits a narrow beam of red light (monochromatic light). Focus the beam of red light on to one side (AB) of the rectangular block as shown in figure 14.6 and mark two points P_1 and P_2 along the incident beam of light with a pencil.

3. Observe the beam that emerges from side (CD) and mark two points P_3 and P_4 along the emergent beam.

4. Remove the glass block and draw a line joining the incident ray and the emergent ray. This is the refracted ray.

5. Draw a normal (perpendicular) at the point of incidence and at the point of emergence of the ray.

6. Measure the angle of incidence *(i)* and angle of refraction *(r)*.

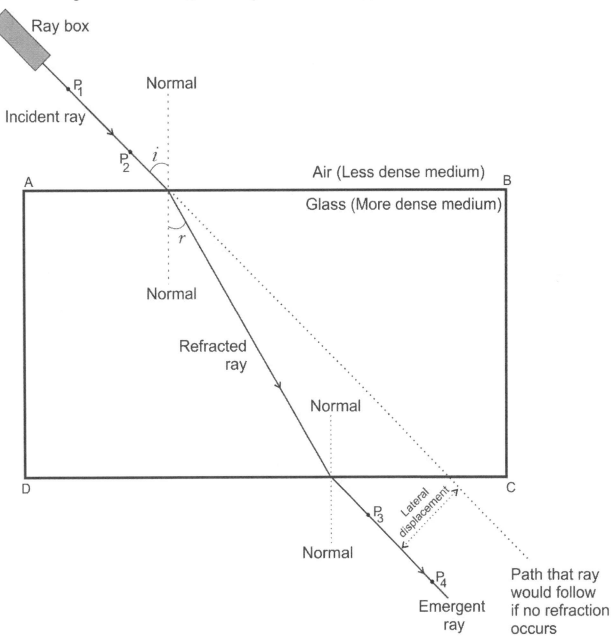

Figure 14.6: Experiment to show refraction of light through a glass block

14 (h, i) Critical angle and total internal reflection

- If a ray of light passes from a denser to a rarer medium then it bends away from the normal, as shown in figure 14.8 when the angle of incidence is 30^0.

- If the angle of incidence in the denser medium is increased to 49^0 as shown in figure 14.8 then the angle between the refracted ray and the normal is 90^0. The angle of incidence at which the angle of refraction is equal to 90^0 is called as the critical angle i_c.

- If the angle of incidence in the denser medium is further increased above the critical angle, then the ray is reflected back into the denser medium following the laws of reflection as shown in figure 14.8 when the angle of incidence is 50^0. This is called as total internal reflection.

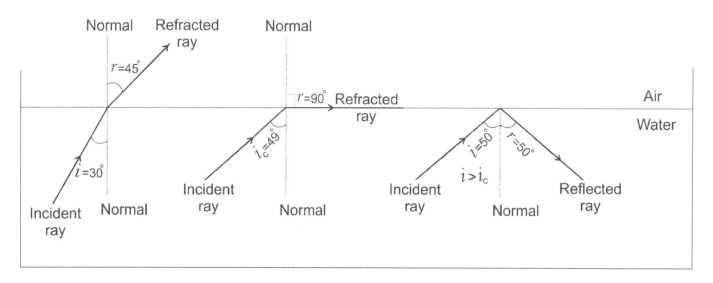

Figure 14.8: Critical angle and total internal reflection

14 (j) Optical fibres in telecommunication

Optical fibres are made of highly refractive glass or plastic. The fibres carry telecommunication signals at the speed of light and can carry much more information than copper wires. They are also cheaper to manufacture than copper wires and are lighter, making it easy to transport.

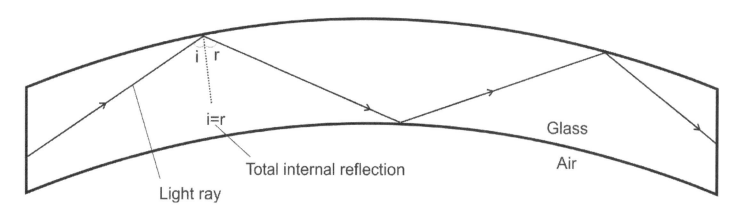

Figure 14.9: Total internal reflection in an optical fibre

14 (i) Experiment to show total internal reflection

Procedure

1.   Place a semi-circular glass block on a white sheet of paper and draw the outline using a sharp pencil.
2. Use a ray box which emits a narrow beam of red light (monochromatic light). Trace the incident ray of light and the refracted ray of light as shown in figure 14.10.
3. The incident ray does not deflect on meeting the glass as it is passing along the normal.
4. Remove the glass block and draw a line joining the incident ray and the refracted ray.

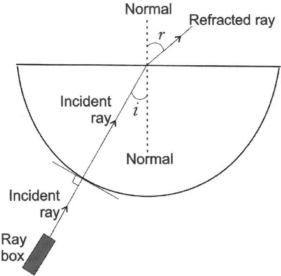

Figure 14.10: Refraction of light

5. Increase the angle of incidence by moving the ray box, until the angle of refraction is 90^0. This angle of incidence is the critical angle(i_c). The critical angle of glass is usually 42^0.

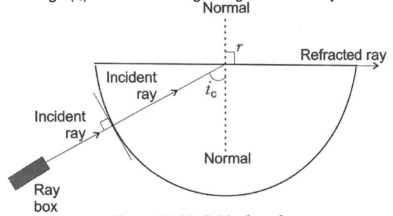

Figure 14.11: Critical angle

6. Increase the angle of incidence by moving the ray box, until the angle of incidence is greater than the critical angle, as shown in figure 14.12. All the light is now reflected back into the glass block. This is called total internal reflection. The angle of incidence is equal to the angle of reflection.

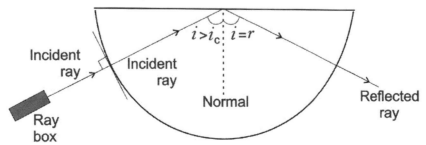

Figure 14.12: Total internal reflection

14 (k, l, m) Refraction by a convex (converging) lens

Example one
If a ray of light passes parallel to the principal axis(AB), then it is refracted to pass through the principal focus (F).

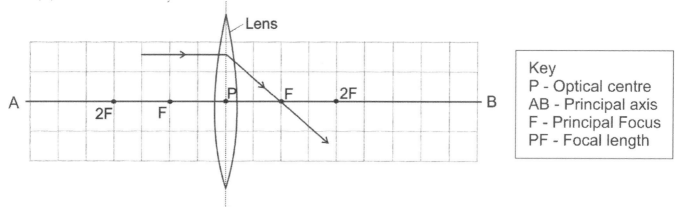

Key
P - Optical centre
AB - Principal axis
F - Principal Focus
PF - Focal length

Example two
If a ray of light passes through the optical centre (P), then it passes straight through without being deflected.

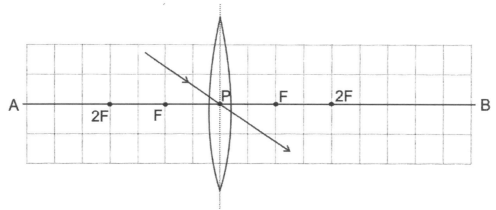

Figure 14.14: Ray passing through the optical centre

Example two
If the incident ray passes through the principal focus and strikes the lens, then the refracted ray is parallel to the principal axis.

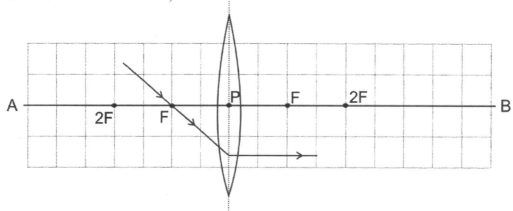

Figure 14.15: Ray passing through principal focus

Ray diagrams to illustrate the formation of real and virtual images of an object by a converging lens

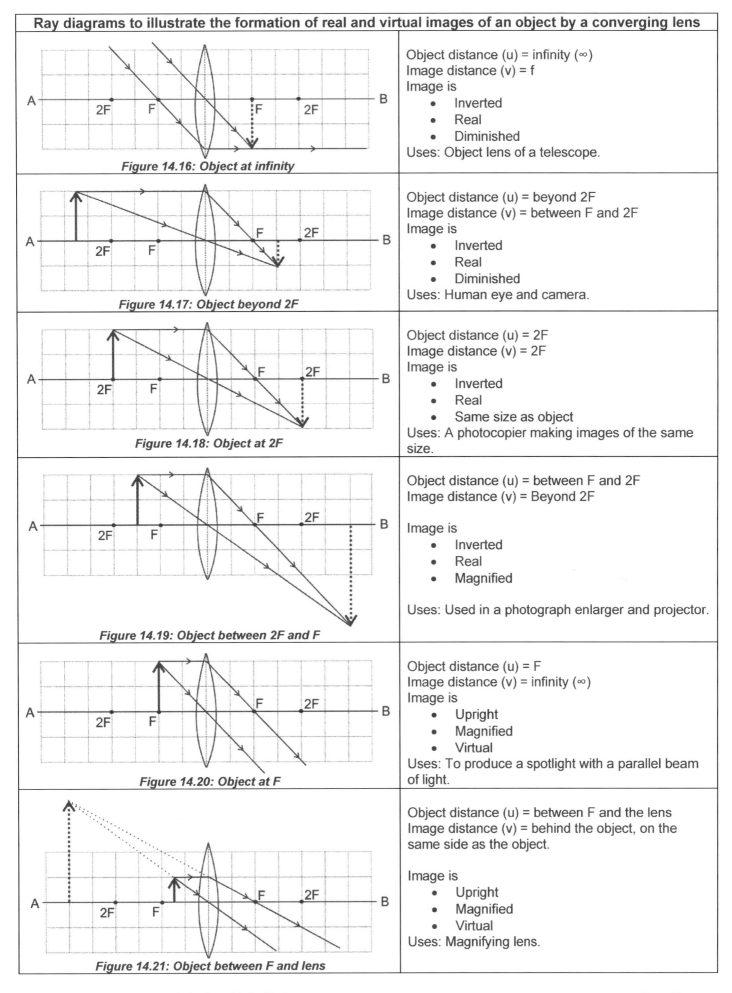

Figure 14.16: Object at infinity

Object distance (u) = infinity (∞)
Image distance (v) = f
Image is
- Inverted
- Real
- Diminished

Uses: Object lens of a telescope.

Figure 14.17: Object beyond 2F

Object distance (u) = beyond 2F
Image distance (v) = between F and 2F
Image is
- Inverted
- Real
- Diminished

Uses: Human eye and camera.

Figure 14.18: Object at 2F

Object distance (u) = 2F
Image distance (v) = 2F
Image is
- Inverted
- Real
- Same size as object

Uses: A photocopier making images of the same size.

Figure 14.19: Object between 2F and F

Object distance (u) = between F and 2F
Image distance (v) = Beyond 2F

Image is
- Inverted
- Real
- Magnified

Uses: Used in a photograph enlarger and projector.

Figure 14.20: Object at F

Object distance (u) = F
Image distance (v) = infinity (∞)
Image is
- Upright
- Magnified
- Virtual

Uses: To produce a spotlight with a parallel beam of light.

Figure 14.21: Object between F and lens

Object distance (u) = between F and the lens
Image distance (v) = behind the object, on the same side as the object.

Image is
- Upright
- Magnified
- Virtual

Uses: Magnifying lens.

Refraction by a concave (diverging) lens

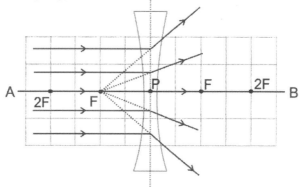

Figure 14.22: Rays passing through the a concave lens

- An incident ray of light that is parallel to the principal axis will bend away from the principal axis after passing through the lens. It will appear to be travelling from the Principal focus, as shown in figure 14.22.
- A ray of light passing through the optical centre remains undeflected after passing through the lens, as shown in figure 14.22.

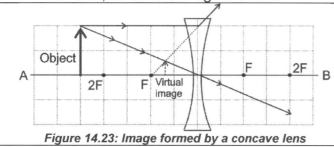

Figure 14.23: Image formed by a concave lens

Image formed by the concave lens shown in figure 14.23 is

- Upright
- Diminished
- Virtual

14 (n) Linear magnification (draw scale diagrams)

Each square in figure 14.24 represents 1 cm. The object distance from the lens is 3 cm. The image distance is 6 cm. The height of the object is 2 cm and the height of the image is 4cm. The image is between F and 2F and the image is formed beyond 2F on the other side of the lens.

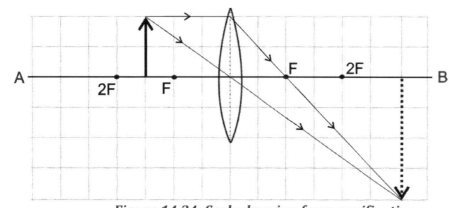

Figure 14.24: Scale drawing for magnification

> **Remember:**
> Magnification = image size/object size
> Or
> Magnification =
> image distance/object distance

Worked example

Using the information in figure 14.24, the magnification can be found as follows.

Magnification = image size/object size

$$= 4 / 2$$

$$= x2$$

Or

Magnification = image distance/object size

$$= 6 / 3$$

$$= x2$$

14 (o) Use of a convex lens in a camera

Figure 14.25 shows the image formed in a camera by using a convex lens. The image is inverted, real and diminished.

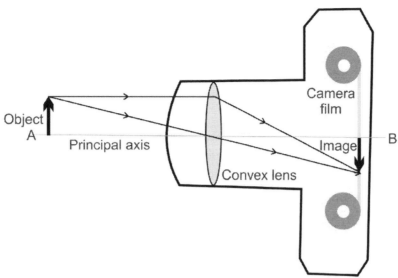

Figure 14.25: Image formed in a camera

Use of lens in a projector and photographic enlarger

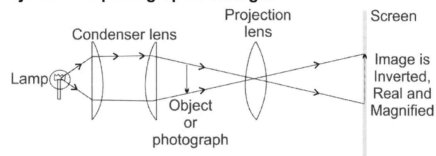

Figure 14.26: Ray diagram of projector and photographic enlarger

14 (p, q) Long sightedness, short sightedness and correction by lenses

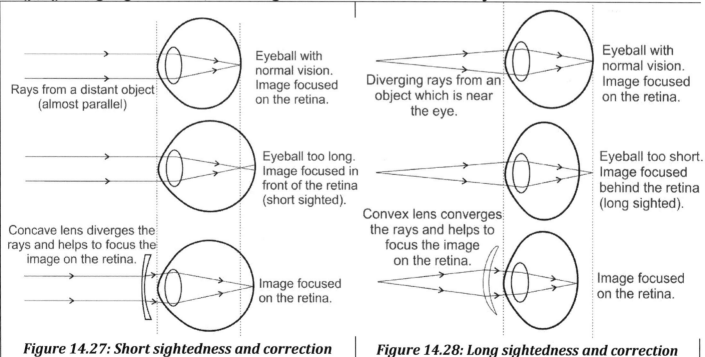

Figure 14.27: Short sightedness and correction | *Figure 14.28: Long sightedness and correction*

Cambridge 5054 syllabus specification 15 (a) *describe the dispersion of light as illustrated by the action on light of a glass prism.*

Cambridge 5054 syllabus specification 15 (b) *state the colours of the spectrum and explain how the colours are related to frequency/wavelength.*

Cambridge 5054 syllabus specification 15 (c) *state that all electromagnetic waves travel with the same high speed in air and state the magnitude of that speed.*

Cambridge 5054 syllabus specification 15 (d) *describe the main components of the electromagnetic spectrum.*

Cambridge 5054 syllabus specification 15 (e) *discuss the role of the following components in the stated applications:*

(1) radio waves – radio and television communications,
(2) microwaves – satellite television and telephone,
(3) infra-red – household electrical appliances, television controllers and intruder alarms,
(4) light – optical fibres in medical uses and telephone,
(5) ultra-violet – sunbeds, fluorescent tubes and sterilisation,
(6) X-rays – hospital use in medical imaging and killing cancerous cells, and engineering applications such as detecting cracks in metal,
(7) gamma rays – medical treatment in killing cancerous cells, and engineering applications such as detecting cracks in metal.

15 (a, b) Dispersion of light

The splitting of white light into its constituent wavelengths is called as dispersion.

The deviation of light is caused due to the refraction or bending of light as it passes through the prism, as shown in figure 15.1. The Violet light deviates the maximum and the red light deviates the minimum. This is dependent on the wavelength of light. Red light has the highest wavelength and lowest frequency and the bending is minimum. On the other hand, violet light has the shortest wavelength and greatest frequency and the deviation is maximum.

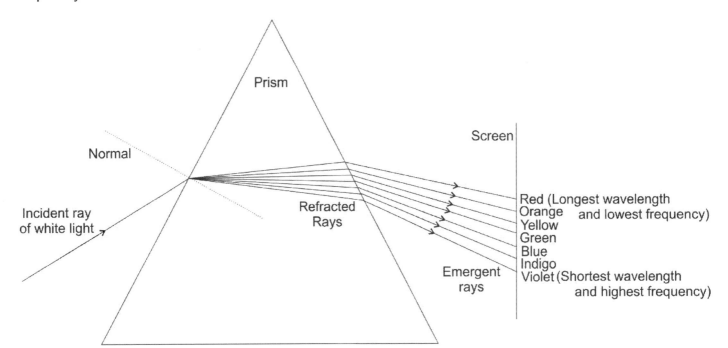

Figure 15.1: Dispersion through a prism

15 (c, d, e) Electromagnetic waves

The electromagnetic spectrum refer to a range of electromagnetic waves with variable wavelength. All electromagnetic waves travel at a speed of 3×10^8 m/s through air and vacuum. The electromagnetic spectrum is shown in figure 15.2.

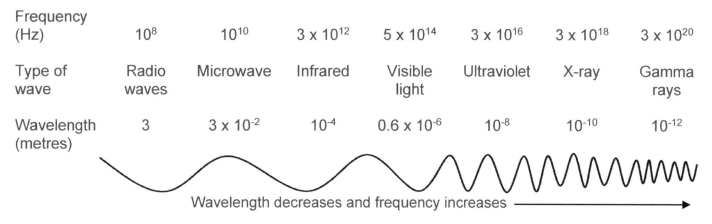

Frequency (Hz)	10^8	10^{10}	3×10^{12}	5×10^{14}	3×10^{16}	3×10^{18}	3×10^{20}
Type of wave	Radio waves	Microwave	Infrared	Visible light	Ultraviolet	X-ray	Gamma rays
Wavelength (metres)	3	3×10^{-2}	10^{-4}	0.6×10^{-6}	10^{-8}	10^{-10}	10^{-12}

Wavelength decreases and frequency increases ⟶

Figure 15.2: Electromagnetic spectrum

Components of the electromagnetic spectrum

Component	Properties	Uses
(1) Radio-waves	Induces alternating currents in metal antenna or aerials.	Used in radio and television communications
(2) Micro-waves	Absorbed by water and fats in food	Used in satellite television, microwave ovens and telephones
(3) Infra-red	Causes heating when absorbed.	Used in household electrical appliances, television controllers and intruder alarms
(4) Light	Refracted by glass lenses and prism	Used in optical fibres in medical uses and telephone
(5) Ultra-violet	Absorbed by glass; causes sunburn; damages and kills living cells.	Used in sunbeds, fluorescent tubes and sterilisation
(6) X-rays	Very penetrating and dangerous.	Hospital use in medical imaging and killing cancerous cells, and engineering applications such as detecting cracks in metal,
(7) Gamma rays	Very penetrating and dangerous.	Medical treatment in killing cancerous cells, and engineering applications such as detecting cracks in metal.

Chapter Sixteen
Sound

Cambridge 5054 syllabus specification 16 (a) *describe the production of sound by vibrating sources.*
Cambridge 5054 syllabus specification 16 (b) *describe the longitudinal nature of sound waves and describe compression and rarefaction.*
Cambridge 5054 syllabus specification 16 (c) *state the approximate range of audible frequencies.*
Cambridge 5054 syllabus specification 16 (d) *explain why a medium is required in order to transmit sound waves and describe an experiment to demonstrate this.*
Cambridge 5054 syllabus specification 16 (e) *describe a direct method for the determination of the speed of sound in air and make the necessary calculation.*
Cambridge 5054 syllabus specification 16 (f) *state the order of magnitude of the speeds of sound in air, liquids and solids.*
Cambridge 5054 syllabus specification 16 (g) *explain how the loudness and pitch of sound waves relate to amplitude and frequency.*
Cambridge 5054 syllabus specification 16 (h) *describe how the reflection of sound may produce an echo.*
Cambridge 5054 syllabus specification 16 (i) *describe how the shape of a sound wave as demonstrated by a cathode-ray oscilloscope (c.r.o.) is affected by the quality (timbre) of the sound wave.*
Cambridge 5054 syllabus specification 16 (j) *define ultrasound.*
Cambridge 5054 syllabus specification 16 (k) *describe the uses of ultrasound in cleaning, quality control and pre-natal scanning.*

16 (a, b, c) Production of sound by vibrating sources

Sound is a form of energy that is produced by a vibrating object. The vibrating object moves back and forth and causes nearby air molecules to vibrate. Also the air molecules that are vibrating influence other air molecules to vibrate. This causes the sound to travel through the air in the form of compressions and rarefactions in longitudinal waves, as shown in figure 16.1.

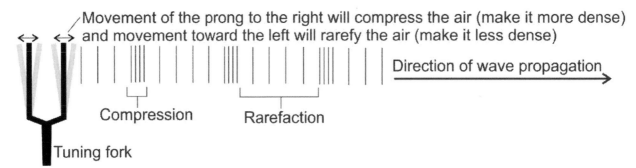

Figure 16.1: Production and transmission of sound

The air molecules around the object will attain the same frequency as the object that is vibrating. The sound always moves away from the source. Humans can hear sound waves that have a frequency between 20Hz to 20,000 Hz. This is called as the audible frequency.

14 (d) Bell jar experiment

Since sound is transmitted by longitudinal waves it needs a medium for transmission. The experiment below shows an experiment to demonstrate that sound cannot travel through a vacuum.

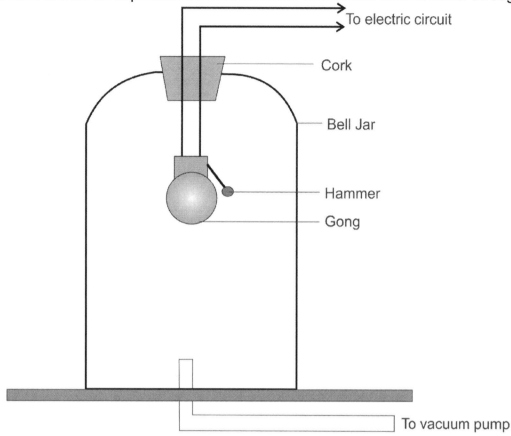

An electric bell is suspended inside an airtight glass bell jar connected to a vacuum pump.

As the electric bell circuit is switched on, the sound is heard. Now if the air is slowly removed from the bell jar by using a vacuum pump, the intensity of sound goes on decreasing and finally no sound is heard when all the air is drawn out.

The hammer keeps striking the gong repeatedly. This clearly proves that sound requires a material for its propagation.

Figure 16.2: Bell jar experiment for transmission of sound

14 (e) Direct Method for determining the speed of sound in air
Manual method

The speed of sound in air is the distance travelled by sound per unit time. The speed of sound in air can be calculated by the direct method, as illustrated in figure 16.3.

- Volunteer A and B stand 100m apart from each other.
- Volunteer B fires a starting pistol.
- Volunteer A starts a stopwatch as soon as he sees the flash of the starting pistol.
- Volunteer A stops the stopwatch as soon as he hears the sound.

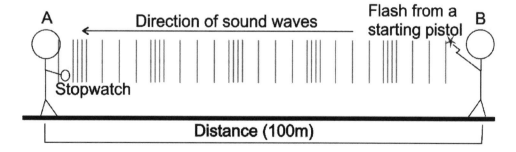

Figure 16.3: Direct method of determining speed of sound manually

Worked example

Using the information in figure 16.3 determine the speed of sound in air if volunteer A records a time of 0.3seconds.

Speed = distance / time
$$= 100/0.3$$
$$= 330 \text{ ms}^{-1}$$

The manual method is less accurate due to reaction time errors. Specially, if the distance between the two volunteers is very low, as the time period will be very low and the reaction time will occupy a large proportion of the total time recorded.

Digital method

Set up the apparatus as shown in figure 16.4.
The timer is switched on as soon as sound from the source reaches microphone A.
The timer is switched off as soon as the sound reaches microphone B.
The distance between the two microphones is measured by a measuring tape.

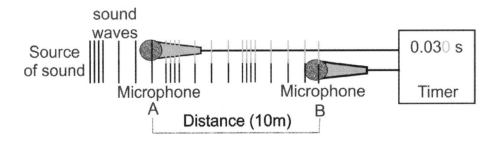

Figure 16.4: Direct method of determining speed of sound digitally

Worked example

Using the information in figure 16.4 determine the speed of sound in air.

Speed = distance / time
$$= 10/0.03$$
$$= 330 \text{ ms}^{-1}$$

14 (f) Speed of sound in air, liquids and solids

The speed of sound in solids is the fastest. Likewise, the speed of sound in gas is the slowest. The order of speed of sound in solids, liquids and gases is shown below.

$$V_{solids} > V_{liquids} > V_{gases}$$

A simple experiment can be used to demonstrate that sound travels more rapidly through solids than through air is shown in figure 16.5.

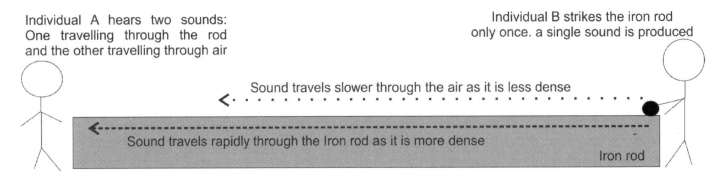

Figure 16.5: Sound travels more rapidly through solids

14 (g) Loudness, pitch, amplitude and frequency of sound waves

If a Cathode Ray Oscilloscope (C.R.O.) is connected to a microphone, wave forms can be generated to understand the properties of sound. A few wave forms and their corresponding relationship with loudness and pitch are described in figure 16.6 and 16.7.

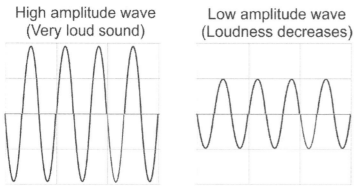

Figure 16.6: Waves with different amplitude

> **Remember:**
>
> ***Loudness α amplitude***
>
> *The figure 16.6 shows two different waves with the same frequency, but with different amplitude. The waves with the larger amplitude will be louder than the waves with the lower amplitude.*

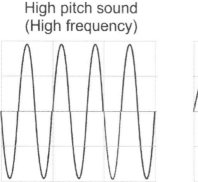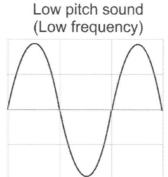

Figure 16.7: Waves with different frequency

> **Remember:**
>
> ***Pitch α Frequency***
>
> *The figure 16.7 shows two different waves with the same amplitude, but with different frequencies. The waves with the larger frequency will have a higher pitch. A shrill note has a high pitch and a coarse note has a low pitch. A girls voice usually has a high pitch and a man's voice has a low pitch.*

14 (h) Reflection of sound may produce an echo

An echo is the sound heard after the reflection of sound from a hard flat surface. If the reflected sound reaches the source at least 0.1 seconds after the first sound is heard then an echo is heard. For that the wall or cliff from which the sound is being reflected should usually be at least 17m away from the source.

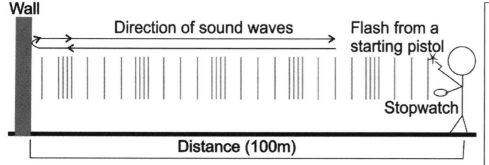

Figure 16.8: Speed of sound by the indirect (echo) method

Figure 16.8 shows how the speed of sound can be calculated by using an echo (indirect method).

Speed of sound = 2d / t
= (2 x 100)/0.6
= 330 m/s

14 (i) Quality (timbre) of sound waves and c.r.o. traces

The figure 16.9 shows a Cathode Ray Oscilloscope (C.R.O.) connected to a microphone. The microphone receives sound waves and converts it into electrical energy. The electrical signals from the microphone are connected to the Y plates of the C.R.O. The pattern of waves formed by the C.R.O. trace is unique to each source. This quality of sound is called timbre.

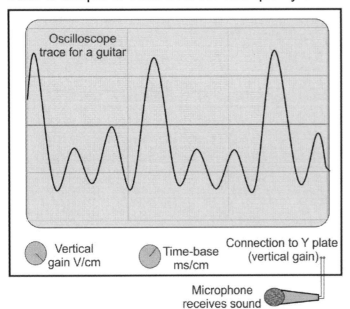

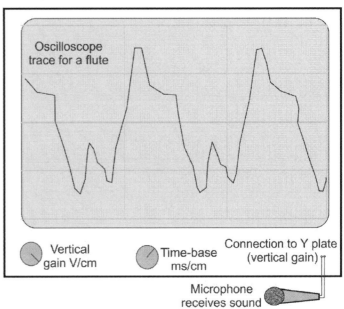

Figure 16.9: Sound waves from two different instruments showing differences in waveform

Figure 16.10 shows a waveform with three fundamental waves AB, BC and CD. Smaller overtones combine with this fundamental wave to produce a unique quality called timbre.

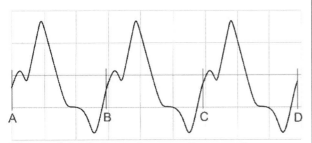

Figure 16.10: Sound waves from a piano

Remember:
The timbre is determined by a fundamental frequency and many smaller frequencies or overtones. The final wave form is a combination of the smaller waveforms and the fundamental wave. Every musical instrument sounds different from the other because it will produce a unique combination of fundamental waves and overtones and the waveforms will have a different shape, even though they may have the same frequency or amplitude. This is called as timbre.

14 (j) Ultrasound

Ultrasound or ultrasonic waves are sound waves with frequencies above 20,000 Hz. These waves are undetectable by humans as they lie above the audible range. Ultrasonic waves have high frequencies and short wavelengths.

14 (k)Uses of Ultrasound

Pre-natal testing

The ultrasonic source sends out vibrations into the tissues. These vibrations cause atoms in the baby's tissues to vibrate and form rarefactions and compressions. The waves travel faster through denser tissue like bones and slower through less dense tissue like muscle. The receiver detects the reflected waves from different tissues at different times and uses this to build a computer generated image of the tissues.

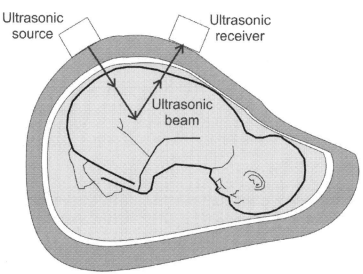

Figure 16.11: Prenatal Ultrasound testing

Worked example

The ultrasound has a wavelength of 1.1 × 10⁻³ m. The speed of the ultrasound in the human soft tissue is 1500 m/s. Calculate the frequency of the ultrasound.

$Velocity = frequency \times wavelength$

$V = f \times \lambda \; or \; f = V / \lambda$

$f = 1500 / (1.1 \times 10^{-3})$

$\quad = 1363636 \; Hz$

$\quad = 1.3 \; MHz$

Ultrasound cleaning

Ultrasonic cleaning is a technology that uses high frequency ultrasound waves to agitate an aqueous or organic medium. The energy is transmitted as waves which acts on dirt that sticks to metals, plastics, glass, rubber and ceramics.

Quality control

Ultrasonic waves can be used in maintenance of railway lines, components of machines and other metals for the detection of cracks.

Ultrasonic sound waves are sent into metals from a source. The wave is reflected back to the source from the boundary of the metal and air. If the metal is intact, the wave takes a longer time to reach the sensor, as shown in figure 16.12. However, if the metal is cracked then the wave travels lesser distance and reaches the sensor earlier. This can be visualized by using a C.R.O with a fixed time base. The waves from the source and sensor will be closer together if the metal is cracked, as shown in figure 16.12.

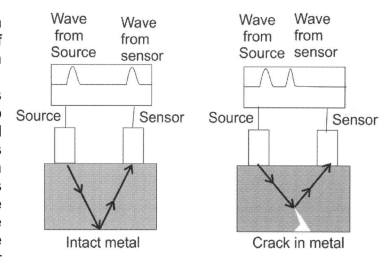

Figure 16.12: Ultrasonic detection of cracks in metals

Chapter Seventeen
Magnetism and Electromagnetism

Cambridge 5054 syllabus specification 17 *(a) state the properties of magnets.*

Cambridge 5054 syllabus specification 17 *(b) describe induced magnetism.*

Cambridge 5054 syllabus specification 17 *(c) state the differences between magnetic, non-magnetic and magnetised materials.*

Cambridge 5054 syllabus specification 17 *(d) describe electrical methods of magnetisation and demagnetisation.*

Cambridge 5054 syllabus specification 17 *(e) describe the plotting of magnetic field lines with a compass.*

Cambridge 5054 syllabus specification 17 *(f) state the differences between the properties of temporary magnets (e.g. iron) and permanent magnets (e.g. steel).*

Cambridge 5054 syllabus specification 17 *(g) describe uses of permanent magnets and electromagnets.*

Cambridge 5054 syllabus specification 17 *(h) explain the choice of material for, and use of, magnetic screening.*

Cambridge 5054 syllabus specification 17 *(i) describe the use of magnetic materials in a computer hard disk drive.*

Cambridge 5054 syllabus specification 17 *(j) describe the pattern of the magnetic field due to currents in straight wires and in solenoids and state the effect on the magnetic field of changing the magnitude and direction of the current.*

Cambridge 5054 syllabus specification 17 *(k) describe applications of the magnetic effect of a current in relays, circuit-breakers and loudspeakers.*

17 (a) Properties of magnets

1. A magnet attracts magnetic materials such as iron, steel, nickel and cobalt.
2. A magnet has two poles, a north pole and a south pole.
3. A freely suspended magnet will come to rest in a geographic North-South direction.
4. Like poles repel and unlike poles attract. Repulsion is a sure test of magnetism.

17 (b) Induced magnetism.

The phenomenon due to which a magnetic substance behaves like a magnet when place near a strong magnet is called magnetic induction.

The poles induced in the soft iron will cause the iron core to be attracted by the magnet.

The poles induced in the soft iron will cause the iron core to be attracted by the magnet.

Figure 17.1: Induced magnetism

17 (c) Magnetic, non-magnetic and magnetised materials

- A magnetic material can be attracted by a magnet.
- Non-magnetic materials cannot be attracted by a magnet.
- Magnetized materials are those materials which exhibit magnetic properties. Magnetisation of a magnetic substance can be brought about by the stroking method or by electromagnetism.

Magnetisation of a magnetic material can be brought about by stroking with another magnet, as shown in figure 17.2 and figure 17.3.

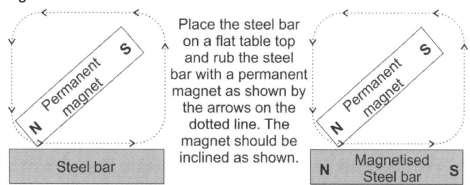

Place the steel bar on a flat table top and rub the steel bar with a permanent magnet as shown by the arrows on the dotted line. The magnet should be inclined as shown.

Figure 17.2: Single stroke method of magnetisation

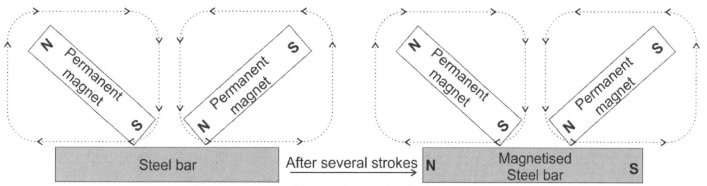

Figure 17.3:Double stroke method of magnetisation

17 (d) Electrical method of magnetisation

An electromagnet can be prepared by placing a steel bar inside a solenoid, as shown in figure 17.4. The solenoid should be wrapped around a cardboard. When a direct current is passed through the solenoid and then switched off, the steel bar is found to be magnetised. The polarity of the magnet can be determined by the clockwise rule as illustrated in figure 17.4.

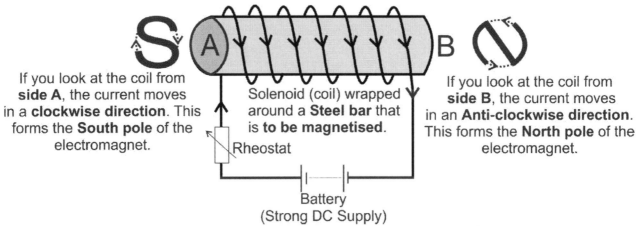

If you look at the coil from **side A**, the current moves in a **clockwise direction**. This forms the **South pole** of the electromagnet.

Solenoid (coil) wrapped around a **Steel bar** that is **to be magnetised**.

Rheostat

Battery
(Strong DC Supply)

If you look at the coil from **side B**, the current moves in an **Anti-clockwise direction**. This forms the **North pole** of the electromagnet.

Figure 17.4: Electrical method of magnetising a steel bar

Electrical method of demagnetisation

A permanent magnet can be demagnetised by placing the magnet inside a cardboard tube surrounded by a solenoid, as shown in figure 17.5. Switch on an alternating current supply and gentlypull the magnet out of the solenoid in a westerly direction. This will demagnetise the magnet.

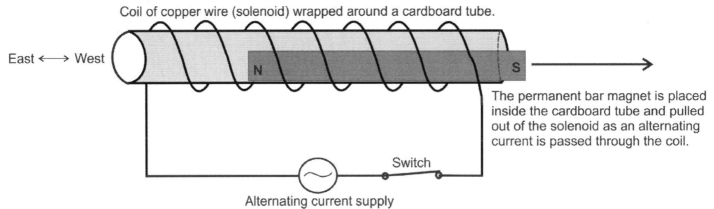

Coil of copper wire (solenoid) wrapped around a cardboard tube.

East ←——→ West

The permanent bar magnet is placed inside the cardboard tube and pulled out of the solenoid as an alternating current is passed through the coil.

Switch

Alternating current supply

Figure 17.5: Electrical method of demagnetisation

17 (e) Plotting of magnetic field lines with a compass

The magnetic field lines indicate the direction in which the North pole of a compass needle will point when placed in a magnetic field.

A compass needle is a small, lightweight magnet that can spin freely. When the needle comes to rest, the North pole gives the direction of the magnetic field. The figure 17.6 shows a plotting compass, which may be used to plot the magnetic lines of force in a magnetic field. The plotting compass is often represented by a circle with an arrow, as shown in figure 17.7. The arrow head indicates the north pole of the plotting compass needle.

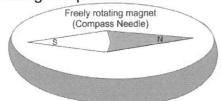

Figure 17.6: A plotting compass

Figure 17.7: Representation of a compass

Figure 17.8 illustrates the procedure for plotting the magnetic lines of force for a bar magnet. The magnetic lines of force
- Always run from North pole towards the south pole.
- Never intersect each other.
- Represents the resultant force of the Earth's magnetic field and the bar magnet on the compass needle.

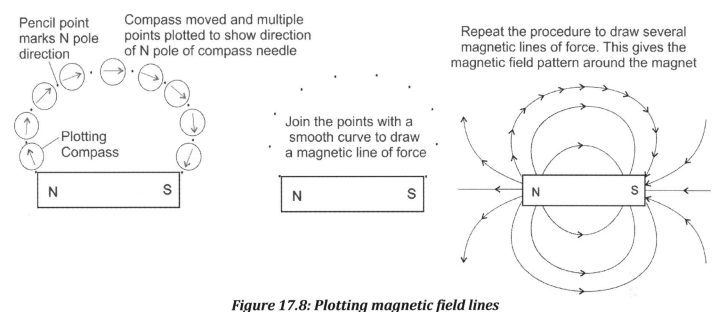

Figure 17.8: Plotting magnetic field lines

Magnetic field pattern between two magnets

The magnetic field pattern around two bar magnets in the Earth's magnetic field is shown in figure 17.9. At a particular point, if the compass needle does not point in any particular direction, then there is no net magnetic field at the point. Such a point is called Neutral point (X) or the Null point. A neutral point is a point where the resultant magnetic field is zero.

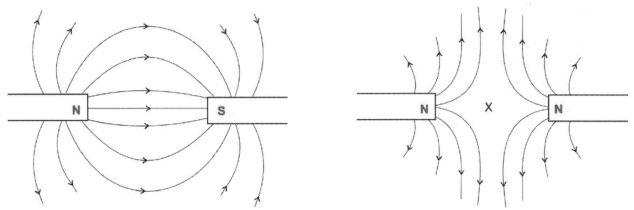

Figure 17.9: Magnetic field between two bar magnets

17 (f) Differences between temporary magnets and permanent magnets

Temporary magnet	Permanent magnet
Usually made from soft iron.	Usually made from steel.
Can be easily magnetized.	Difficult to magnetise.
Can be easily demagnetised.	Difficult to demagnetise.

17 (g) Uses of permanent magnets and electromagnets

Permanent magnets and electromagnets are used In a variety of appliances. Motors, loudspeakers, generators, dynamos, speedometers, medical scanners, maglev trains, calling bells, etc. The use of electromagnets in the electric bell is shown in figure 17.10 and 17.11.

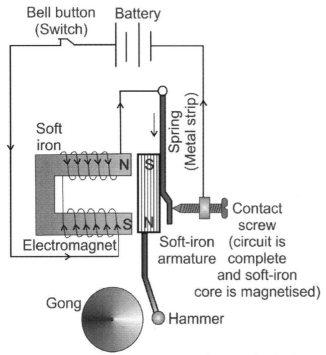

Figure 17.10: Circuit complete in the bell

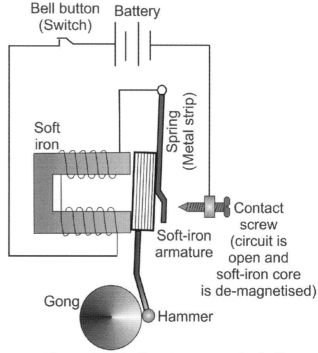

Figure 17.11: Circuit open in the bell

When the switch is on, the circuit is complete. Current passes through the coil and the soft iron gets magnetised. This induces magnetism in the soft iron armature and the armature gets attracted to the soft iron electromagnet. The hammer strikes the gong.

When the hammer strikes the gong, the circuit is opened. No current flows through the coil and the soft iron core is demagnetised. The soft-iron armature also gets demagnetised and the spring pulls the armature back to the original position. The cycle is then repeated.

Magnetic effect of current in relays

The relay consists of two circuits. A low voltage Circuit is a simple electromagnet which requires only a small current. When the switch is closed, current flows and the iron rocker arm is attracted to the electromagnet. The arm rotates about the central pivot and pushes the contacts together. The high voltage circuit is now switched on, as shown in figure 17.12.

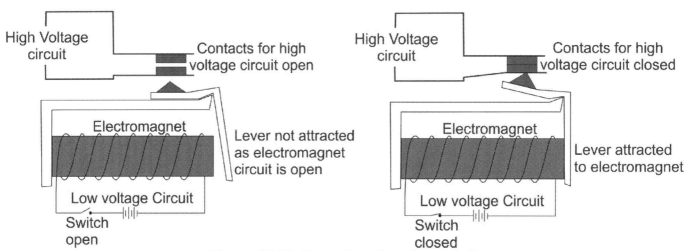

Figure 17.12: Operation of a relay switch

The advantage of using a relay is that a small current can be used to switch on and off a circuit with a large current. This is useful for two reasons.
1. The low current circuit may contain a component such as an LDR, which only uses small currents.
2. Only the high current circuit needs to be made from thick wire.
A relay is used to operate the starter motor in cars and the heating circuit in diesel engines.

17 (k) Magnetic effect of current in Circuit breakers

The circuit breaker is a safety device just like a fuse. Circuit breakers that are used at the distribution board in houses are called MCBs (miniature circuit breakers). The MCBs disconnect the supply if too large a current flows. When the live wire carries the usual operating current, the electromagnet is not strong enough to separate the contacts. If there is a short-circuit in the appliance and the current increases, the electromagnet will pull hard enough to separate the contacts and break the circuit. The spring then keeps the contacts apart. After the fault is repaired, the contacts can be pushed back together by pushing the reset button of the circuit breaker, as shown in figure 17.13.

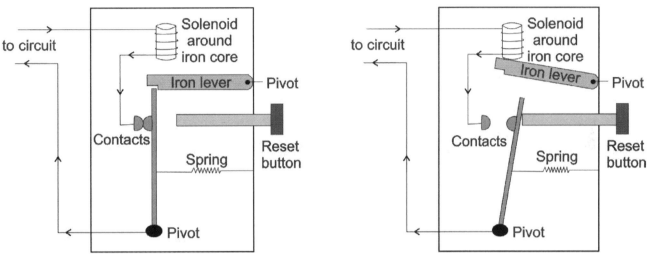

Figure 17.13: Miniature circuit breaker

Magnetic effect of current in loudspeakers

The loudspeaker uses a coil which can slide backwards and forwards over the central pole of a circular permanent magnet. The coil is joined to a paper cone, shown in figure 17.14. The input is an alternating current which makes the coil (and the paper cone) move backwards and forwards at the same frequency as the changing current. The paper cone then moves the air backwards and forwards which creates the sound.

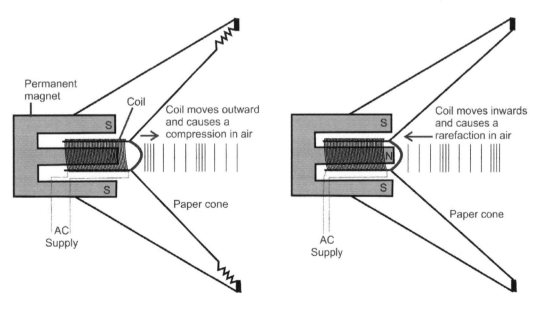

Figure 17.14: Working of a loudspeaker

17 (h) Magnetic Screening

Some devices cannot function properly if there is a magnetic field passing through it. To prevent magnetic fields from disturbing a sensitive device, the device is enclosed in a magnetic material like soft iron or mu-metal (an alloy of nickel and iron). The magnetic material will guide the field away from the sensitive device, as shown in figure 17.15. This is called as magnetic screening or magnetic shielding. Mu-metal is used in electric transformers to prevent the field from affecting other devices, in computer hard drives to keep out magnetic field, and in magnetic resonance imaging.

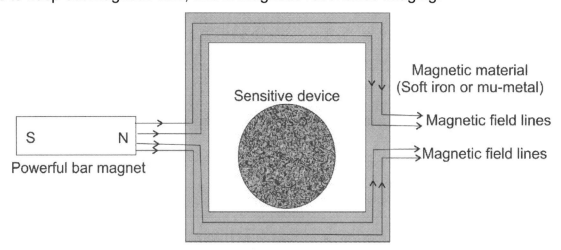

Figure 17.15: Magnetic screening

17 (i) Use of magnetic materials in audio and video tapes

Audio and video tapes are made of a plastic film coated with a magnetic material that has an iron-containing compound coated on its surface. Information can be stored by magnetizing the magnetic material with an electromagnet. The images or audio are built up by magnetizing the tapes to different strengths by using currents of different strength. Care must be taken to store audio and video tapes away from strong magnetic fields as the information stored on these tapes could be wiped out by these strong magnetic fields.

17 (j) Magnetic field due to current in solenoids

A solenoid is a coil which has a smaller diameter than its length. In simple terms it is a long thin coil. When a current passes through a solenoid, a magnetic field is generated around the coil. The pattern of the field is shown in figure 17.16.

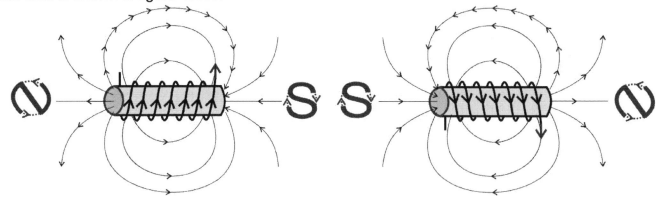

Figure 17.16: Magnetic field around a solenoid

> **Remember:**
> *Reversing the current causes the magnetic field to be reversed. Increasing the strength of the current will increase the strength of the magnetic field. Refer to chapter 23 for more information on determining the polarity of the electromagnet.*

Chapter Eighteen
Static Electricity

18 (b) Nature of electric charges

There are two types of charges in an object: **positive** charges and **negative** charges.

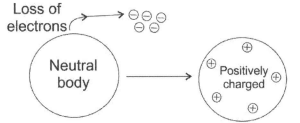

Figure 18.1: A body can become positively charged by losing electrons

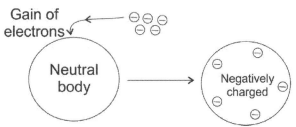

Figure 18.2: A body can become negatively charged by gaining electrons

Note that positive charges never leave or enter a body. It is ALWAYS the flow of NEGATIVE CHARGES (ELECTRONS) that are responsible for charging a body. The positive charges are held in the nucleus of atoms and cannot travel from one body to another.

18 (a) Experiments to show electrostatic charging by friction

Rub a Polythene rod with wool. Electrons flow from the wool to the polythene, as shown in figure 18.3.

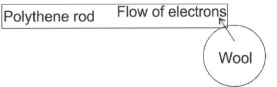

Figure 18.3: electrons flow from wool to polythene

The polythene rod gets negatively charged and the wool gets positively charged, as shown in figure 18.4.

Figure 18.4: polythene is negatively charged and wool is positively charged

Rub a Perspex rod with wool. Electrons flow from the Perspex to the wool, as shown in figure 18.5.

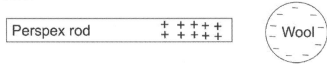

Figure 18.5: electrons flow from Perspex to wool

The Perspex rod gets positively charged and the wool gets negatively charged, as shown in figure 18.6.

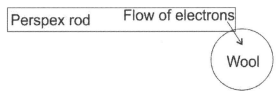

Figure 18.6: Perspex is positively charged and wool is negatively charged.

18 (d) Unlike charges attract and like charges repel, as shown in figure 18.7.

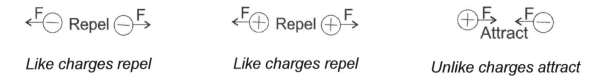

Like charges repel   Like charges repel   Unlike charges attract

Figure 18.7: Like charges repel and unlike charges attract

So, a positively charged Perspex rod will attract a negatively charged polythene rod, as shown in figure 18.8

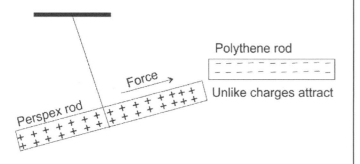

Figure 18.8: bodies with unlike charges attract

Likewise, a negatively charged Polythene rod will repel a negatively charged polythene rod, as shown in figure 18.9

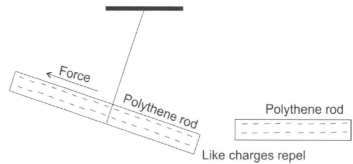

Figure 18.9: bodies with like charges repel

18 (e, f) Electric field

An electric field is a region in which an electric charge experiences a force. The direction in which a small freely moving positive charge would move if placed in an electric field is called an electric line of force.

The direction of lines of force around various objects is shown in figure 18.10. Use the fact that a positive charge will be repelled by another positive charge and will be attracted towards a negative charge.

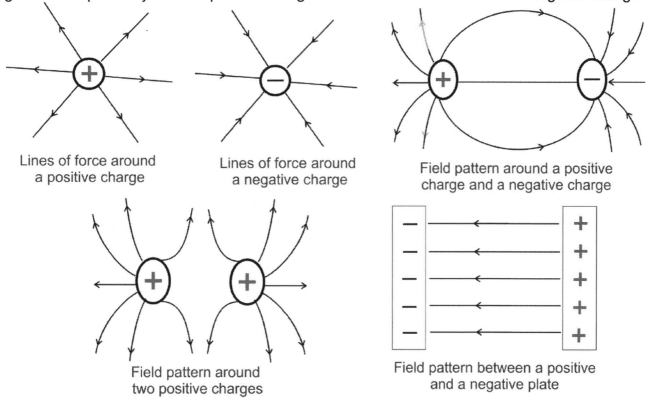

Figure 18.10: Electric field lines and patterns

18 (g) Charging a body by induction
Step 1: Separation of charges by induction, as shown in figure 18.11 and 18.12.

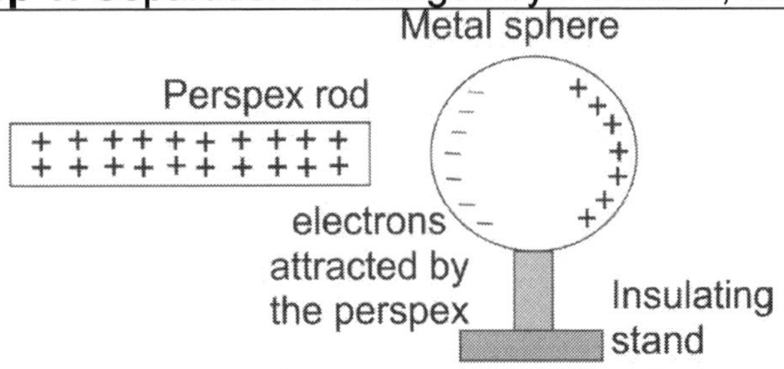

Metal sphere

Perspex rod

+ + + + + + + + +
+ + + + + + + + +

electrons
attracted by
the perspex

Insulating
stand

Figure 18.11: Electrons attracted to Perspex

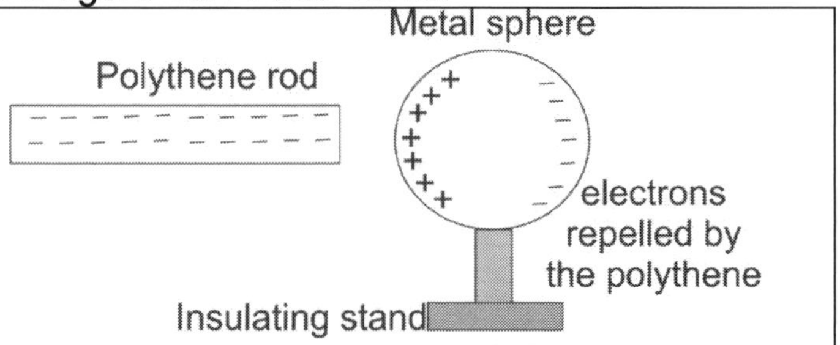

Metal sphere

Polythene rod

- - - - - - - - - -

electrons
repelled by
the polythene

Insulating stand

Figure 18.12: Electrons repelled by Polythene

Step 2: Earth the sphere.

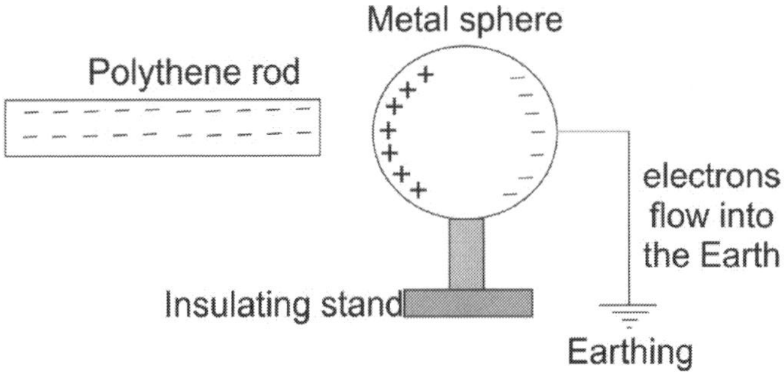

Metal sphere

Polythene rod

- - - - - - - - - -

electrons
flow into
the Earth

Insulating stand

Earthing

Figure 18.13: Electrons flow away from sphere

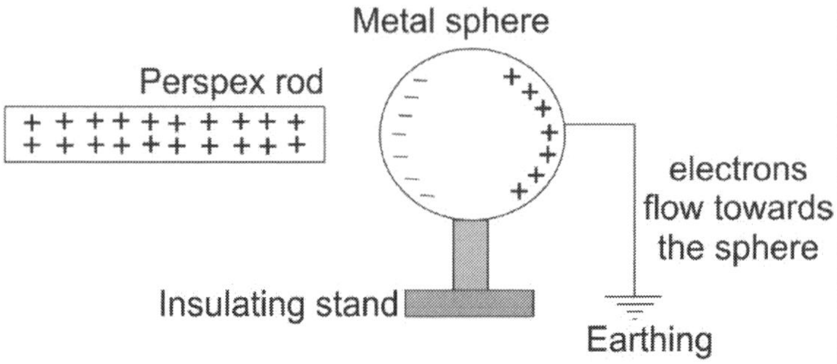

Metal sphere

Perspex rod

+ + + + + + + + +
+ + + + + + + + +

electrons
flow towards
the sphere

Insulating stand

Earthing

Figure 18.14: Electrons flow towards the sphere

Step 3: Disconnect the earthing

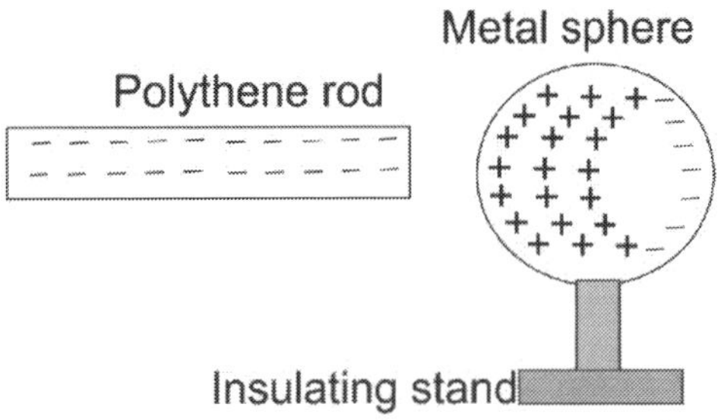

Metal sphere

Polythene rod

- - - - - - - - - -

Insulating stand

Figure 18.15: Sphere more positive charges

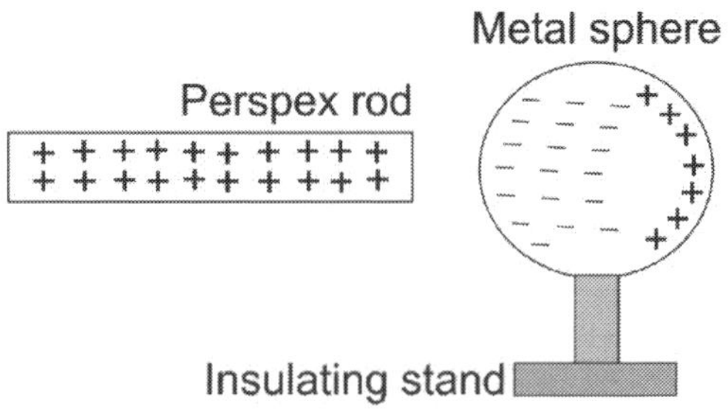

Metal sphere

Perspex rod

+ + + + + + + + +
+ + + + + + + + +

Insulating stand

Figure 18.16: Sphere more negative charges

Step 4: Remove the charged rod

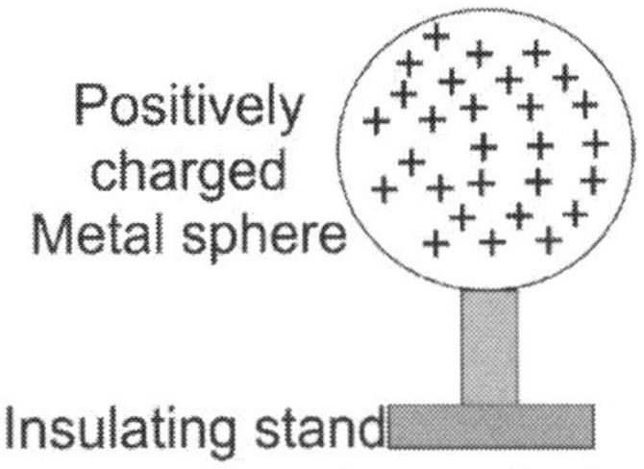

Positively
charged
Metal sphere

Insulating stand

Figure 18.17: Sphere is positively charged

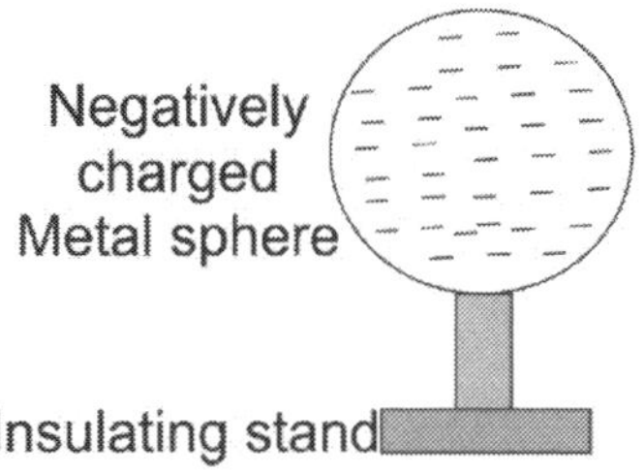

Negatively
charged
Metal sphere

Insulating stand

Figure 18.18: Sphere is negatively charged

18 (i) Earthing a charged object
The earth is an infinite source and sink of electrons.

If a positively charged body is connected to the earth (earthed) then electrons will flow from the earth towards the body and neutralize it.

Likewise, if a negatively charged body is connected to the earth (earthed) then electrons will flow from the body towards the earth.

18 (c) Charge

Charge is measured in coulombs. It is represented by the symbol Q. A charge of 1 coulomb is equal to 6.24×10^{18} electrons.

$$\text{Charge (Q)} = \text{Current (I)} \times \text{Time (t)}$$
$$Q = I \times t$$

18 (j) Lightning and lightning conductors

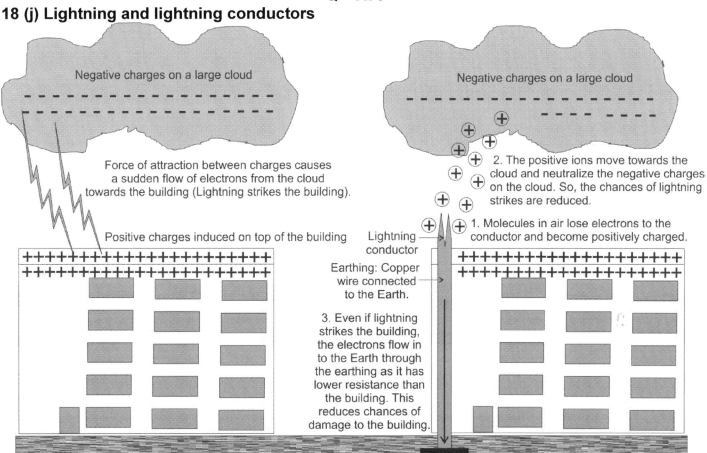

Figure 18.19: Lightning conductor

18 (k) Use of charges in a photocopier

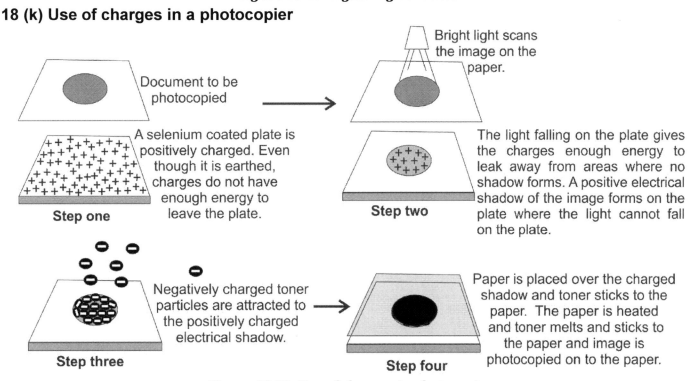

Figure 18.20: Use of charges in photocopiers

18 (k) Use of charges in an electrostatic precipitator

Figure 18.21 illustrates the use of charges in an electrostatic precipitator to remove soot from smoke.

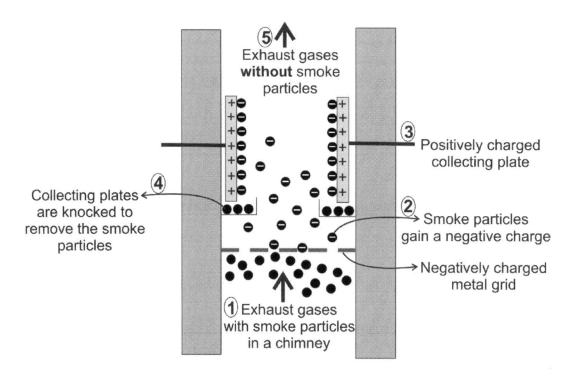

Figure 18.21: Electrostatic precipitator

18 (h) Electrical conductors and insulators

Conductors are materials that will allow charges to pass through them easily. These materials are known as electrical **conductors**.

Examples of conductors are **metals**, such as copper, iron and steel.

Insulators are materials that will not allow charges to pass through them. These materials are known as electrical **insulators**.

Examples of insulators are Plastic, wood, glass and rubber.

Chapter nineteen
Current Electricity

Cambridge 5054 syllabus specification 19 (a) state that a current is a flow of charge and that current is measured in amperes.

Cambridge 5054 syllabus specification 19 (b) recall and use the equation charge = current × time.

Cambridge 5054 syllabus specification 19 (c) describe the use of an ammeter with different ranges.

Cambridge 5054 syllabus specification 19 (d) explain that electromotive force (e.m.f.) is measured by the energy dissipated by a source in driving a unit charge around a complete circuit.

Cambridge 5054 syllabus specification 19 (e) state that e.m.f. is work done/charge.

Cambridge 5054 syllabus specification 19 (f) state that the volt is given by J / C.

Cambridge 5054 syllabus specification 19 (g) calculate the total e.m.f. where several sources are arranged in series and discuss how this is used in the design of batteries.

Cambridge 5054 syllabus specification 19 (h) discuss the advantage of making a battery from several equal voltage sources of e.m.f. arranged in parallel.

Cambridge 5054 syllabus specification 19 (i) state that the potential difference (p.d.) across a circuit component is measured in volts.

Cambridge 5054 syllabus specification 19 (j) state that the p.d. across a component in a circuit is given by the work done in the component/charge passed through the component.

Cambridge 5054 syllabus specification 19 (k) describe the use of a voltmeter with different ranges.

Cambridge 5054 syllabus specification 19 (l) state that resistance = p.d./current and use the equation resistance = voltage/current in calculations.

Cambridge 5054 syllabus specification 19 (m) describe an experiment to measure the resistance of a metallic conductor using a voltmeter and an ammeter and make the necessary calculations.

Cambridge 5054 syllabus specification 19 (n) state Ohm's Law and discuss the temperature limitation on Ohm's Law.

Cambridge 5054 syllabus specification 19 (o) *use quantitatively the proportionality between resistance and the length and the cross-sectional area of a wire.

Cambridge 5054 syllabus specification 19 (p) calculate the net effect of a number of resistors in series and in parallel.

Cambridge 5054 syllabus specification 19 (q) describe the effect of temperature increase on the resistance of a resistor and a filament lamp and draw the respective sketch graphs of current/voltage.

Cambridge 5054 syllabus specification 19 (r) describe the operation of a light-dependent resistor.

19 (a) Current

Current is the rate of flow of charge through an electrical conductor. It is measured in amperes (A) by using an ammeter. The symbol is I. When a charge of 1 Coulomb travels through a conductor in 1 second then the current is 1 ampere.

$$\text{Current (I)} = \frac{Charge(Q)}{time(t)}$$

Where,
I = current in ampere (A)
Q = charge in coulomb (C)
t = time is seconds (s)

Current always flows from the positive to the negative terminal of a D.C. circuit, as shown in figure 19.1. The flow of electrons is always from the negative to the positive terminal. Positive charges in a solid conductor do not move, as they are trapped in heavy nuclei. Before the discovery of the nuclear atom it was thought that current is the flow of positive charges. However, we still refer to the direction in which positive charges may have moved, if they could, as the direction of a conventional current.

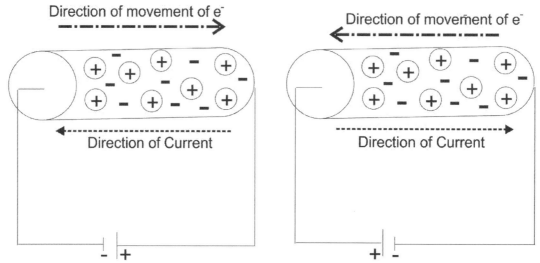

Figure 19.1: Showing a simple circuit and an illustration of the inside of a conductor.

19 (b) Worked example

A current of 3.5 A flows through a conductor. Calculate the charge passing through a point in the conductor in one minute. Show your working. (3)

$Q = I \times t$
$= 3.5 \times 60$
$= 210\ C$

Calculate the time needed for 2 C of charge to flow through a conductor, when a current of 4A passes through the conductor. Show your working. (3)

$t = Q / I$
$= 2 / 4$
$= 0.5\ s$

A student is told that a current of 4 A is passing through the wire. Explain what is meant by this statement. (2)

4A means that a charge of 4C passes through the conductor in 1 second.

Or

4A means that a charge of 8C passes through the conductor in 2 second.

Use the information in Fig. 19.1 to explain why a current flows through the circuit. (4)

The positive terminal of the battery attracts the electrons and the negative terminal repels electrons.

The current in a circuit can be measured by an instrument called the ammeter. An ammeter has a red positive terminal. When connecting an ammeter to a circuit, it should always be connected in series with the circuit. The positive terminal of the ammeter must be connected to the positive terminal of the cell to prevent damage to the ammeter.

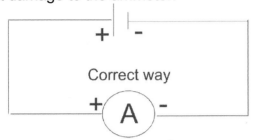

Figure 19.2: Correct way of connecting an ammeter

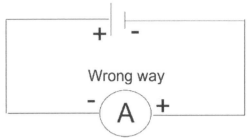

Figure 19.3: Wrong way of connecting an ammeter

19 (c) Sometimes it may be necessary to detect and measure very small currents. A milliammeter or a microammeter may be used.

$1 A = 10^3$ mA (milliamperes)
$1 A = 10^6$ µA (microamperes)

Using a milliammeter to measure a small current will increase the accuracy and precision of the readings.

Worked example

The current through a car headlamp is 200 mA. Express this in terms of amperes. (2)

$1A = 1000mA$
(cross multiply)
$X = 200 mA$

$1000X = 200$
$X = 200/1000$
$= 0.2 A$

The current through a pocket calculator is 0.005 A. Calculate the current in µA. (2)

$1A = 1000000µA$
(cross multiply)
$0.005A = X mA$

$1X = 0.005 X 1000000$
$X = 5000 µA$

The tables below show the ammeters that are available in the laboratory. State which ammeter you would choose to measure the current in a car headlamp. Give reasons for your choice. (3)

Range	Ammeter	Range	Ammeter
0-1 mA	A	0-500 mA	K
0-3 mA	B	0-1 Amps	L
0-5 mA	C	0-3 Amps	M
0-10 mA	D	0-5 Amps	N
0-25 mA	E	0-10 Amps	O
0-50 mA	F	0-15 Amps	P
0-100 mA	G	0-25 Amps	Q
0-150 mA	H	0-30 Amps	R
0-200 mA	I	0-50 Amps	S
0-300 mA	J		

J would be suitable for measuring the current in the car headlamp, as 200mA lies between the range of ammeter J. The maximum voltage that can be measured by ammeters A to I will be too low and the other ammeters will not give sensitive readings. Any deflection in the needle will be too small to read accurately.

19 (d, e, f) Electromotive force (e.m.f.)

The electromotive force (e.m.f.) is measured by the **energy** given out by a source (cell) in moving **one coulomb of charge** around a complete circuit. The electromotive force is measured across the source of electricity (**cell or battery**). It is measured in volts.

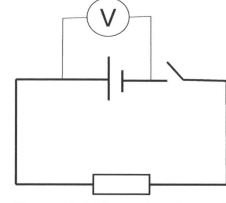

Figure 19.4: Measuring the e.m.f

$$V = \frac{E}{Q} \quad \text{or} \quad V = \frac{W}{Q}$$

Where,
V = Electromotive force of the cell in volts (V)
E = Energy given out by the source or work done in moving the charge in Joules (J)
Q = Charge in coulombs (C)

So, an e.m.f. of 5V means that 5J of work needs to be done to drive 1C of charge around the circuit.

Or

Each coulomb of charge that is driven out of the source possesses 5J of energy.

The electromotive force of a cell is measured by connecting a voltmeter across the terminals of a cell when it is not supplying a current, as shown in figure 19.4.

Worked example

A cell does 3J of work to drive 0.2C of charge around a circuit. Calculate the e.m.f. of the cell.
$E = W / Q$
$= 3 / 0.2$
$= 15\ V$

19 (g, h) Batteries

When several cells are connected together, it is called as a battery. Cells may be connected in series or in parallel, as shown in figure 19.5 and 19.6 respectively.

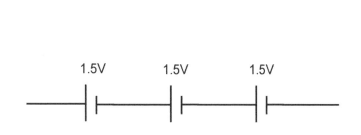

Figure 19.5: Cells in series

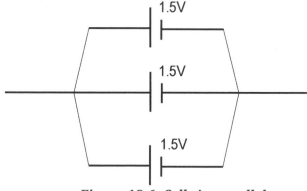

Figure 19.6: Cells in parallel

Remember:
Connecting cells in series
- gives an increase in e.m.f., but,
- get discharged very quickly.
The e.m.f. of the battery in figure 19.5 is 4.5 V

Remember:
Connecting cells in parallel
- gives a decrease in e.m.f., but,
- last longer before getting discharged.
The e.m.f. of the battery is in figure 19.6 is 1.5V

19 (i, j, k) Potential difference

The work done across a conductor (bulb or other circuit component) in moving one coulomb of charge through it is called the potential difference. It may also be defined as the energy lost as one coulomb of charge flows through the component in a circuit.

$$V = \frac{W}{Q}$$

Where,
V = Potential difference or voltage in volts (V)
W = work done in Joules (J)
Q = Charge in coulombs (C)

The potential difference across a component in the circuit can be measured by connecting a Voltmeter in parallel across the circuit component as shown in figure 19.7.

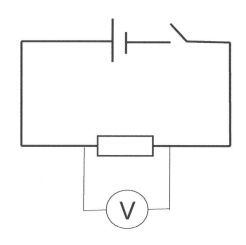

Figure 19.7: Measuring potential difference

19 (l) Resistance

Resistance is the ratio of the potential difference and the current in a circuit. It is measured in ohms (Ω). The resistance is calculated by one of the following equations.

$$R = \frac{V}{I}$$

Where,
R = the resistance in ohm (Ω)
V = the potential difference in volts (V)
I = the current in amperes (A)

Or

$$R = \frac{\rho L}{A}$$

Where,
R = resistance in ohm (Ω)
ρ = resistivity in ohm metre (Ωm)
L = length in metre (m)
A = Area of cross section (m^2)

19 (m) Experiment to measure the resistance of a metallic conductor

The circuit set up in figure 19.8 is used to test the resistance of a metal conductor. The rheostat is used to change the current and voltage in the circuit. The current (I) in the ammeter and the potential difference (V) across the conductor is recorded. The resistance of the conductor can then be calculated by using the expression

$$R = \frac{V}{I}$$

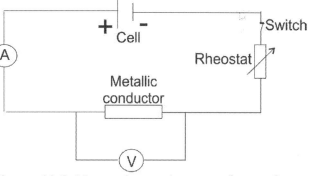

Figure 19.8: Measuring resistance of a conductor

19 (n) Ohm's Law

The potential difference (voltage) across a conductor is proportional to the current passing through it. The ratio of Voltage and current is a constant value called the resistance. Ohm's Law is given by the expression,

$$R = \frac{V}{I}$$

The table gives typical values that may be obtained for a conductor that obeys Ohm's law. The resistance of the conductor is 2 Ω. A graph for Voltage against current will give a straight line.

Voltage (V)	2	4	6	8	10
Current (I)	1	2	3	4	5
Resistance (R) = V / I	2	2	2	2	2

19 (q) Temperature limitation on Ohm's Law

Ohmic conductors obey Ohm's law and have a constant resistance. Non-Ohmic conductors have a variable resistance. The resistance of thermistors decreases as temperature increases. On the other hand the resistance of a filament in a lamp will increase with temperature. So, they do not obey Ohm's law, as depicted in figure 19.9. **Note that the shape of the graphs B and C will change if the axes are reversed.**

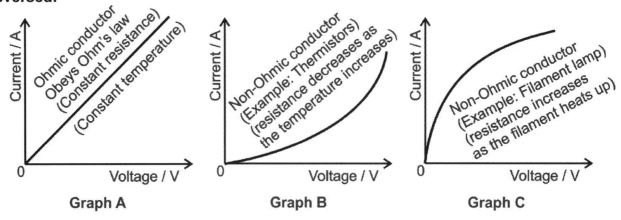

Figure 19.9: Thermistors and filament lamps do not obey Ohm's law due to changes in temperature

19 (o) Proportionality between resistance and the length and the cross-sectional area of a wire

The length of a wire, the cross section and resistance are quantitatively related by the equation

$$R = \frac{\rho L}{A}$$

Where,
R = resistance in ohm (Ω)
ρ = resistivity in ohm metre (Ωm)
L = length in metre (m)
A = Area of cross section (m^2)

Worked examples

A copper wire of resistivity 1.7×10^{-8} Ωm has a length of 100m and a cross section area of $0.00012m^2$. Calculate the resistance of the wire.

$R = \rho L / A$

$= ((1.7 \times 10^{-8}) \times 100) / 0.00012$

$= 0.014\ \Omega$

The same copper wire is cut in half. Calculate the resistance.

$R = \rho L / A$

$= ((1.7 \times 10^{-8}) \times 50) / 0.00012$

$= 0.007\ \Omega$

(Decreasing the length by 50% also decreased the resistance by 50%. So, resistance and length are directly proportional.)

A copper wire of resistivity 1.7×10^{-8} Ωm has a length of 100m and a cross section area of $0.00024m^2$. Calculate the resistance of the wire.

$R = \rho L / A$

$= ((1.7 \times 10^{-8}) \times 100) / 0.00024$

$= 0.007\ \Omega$

(Doubling the cross section area (from $0.00012m^2$ to $0.0024m^2$) has decreased the resistance by half. So, cross section area and resistance are inversely proportional)

Sketch graphs to show the relation between the cross sectional area, length of the conductor and resistance.

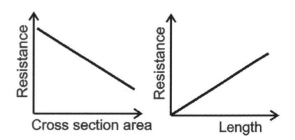

19 (p) Resistors in series and in parallel

Resistors may be connected to in a circuit either in series or parallel. In a series connection, one end of a resistor is connected to the beginning of another resistor. In a parallel connection, the ends of the resistors are connected to common points (A and B) in the circuit, as shown in figure 19.10.

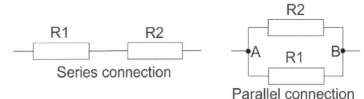

Figure 19.10: Connecting resistors

The combined resistance of resistors in series is

$$R = R1 + R2$$

The combined resistance of resistors in parallel is

$$\frac{1}{R} = \frac{1}{R1} + \frac{1}{R2}$$

Worked examples

Calculate the effective resistance of the combination of resistors shown below.

$\mathcal{R} = \mathcal{R}_1 + \mathcal{R}_2$

$= 2 + 4$

$= 6 \ \Omega$

Calculate the effective resistance of the combination of resistors shown below.

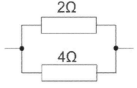

$1/\mathcal{R} = 1/\mathcal{R}_1 + 1/\mathcal{R}_2$

$= 1/4 + 1/2$

$= ((1 \times 1) + (2 \times 1))/4$

$1/\mathcal{R} = 3/4$

$\mathcal{R} = 4/3$

$= 1.33 \ \Omega$

19 (r) Light-dependent resistor (LDR)

A light dependent resistor (LDR) is a component whose resistance changes when the amount of light falling on it (called the light intensity) changes. The resistance decreases as the light intensity increases. An LDR can be used in a circuit to provide an input about changes in the light intensity of the surroundings.

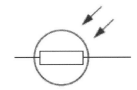

Figure 19.11: Symbol of LDR

Cambridge 5054 syllabus specification 20 (a) **draw circuit diagrams with power sources (cell, battery or a.c. mains), switches (closed and open), resistors (fixed and variable), light-dependent resistors, thermistors, lamps, ammeters, voltmeters, magnetising coils, bells, fuses, relays, diodes and light-emitting diodes.*

Cambridge 5054 syllabus specification 20 (b) *state that the current at every point in a series circuit is the same, and use this in calculations.*

Cambridge 5054 syllabus specification 20 (c) *state that the sum of the potential differences in a series circuit is equal to the potential difference across the whole circuit and use this in calculations.*

Cambridge 5054 syllabus specification 20 (d) *state that the current from the source is the sum of the currents in the separate branches of a parallel circuit.*

Cambridge 5054 syllabus specification 20 (e) *do calculations on the whole circuit, recalling and using formulae including R = V/ I and those for potential differences in series, resistors in series and resistors in parallel.*

20 (a) Symbols used in circuit diagrams

Circuit diagrams are often used to depict the connections between various devices. The symbols in figure 20.1 are used to represent all the appliances that the syllabus has specified. It is necessary to memorise these symbols for use in circuit diagrams.

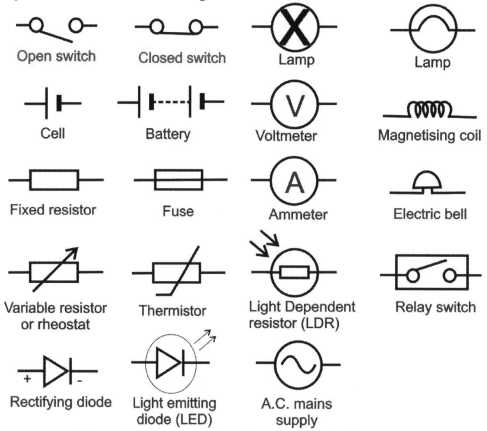

Figure 20.1: Symbols for drawing circuit diagrams

20 (b, c, d, e) Worked examples

Use the circuit rules to answer the questions that follow about the circuit below.

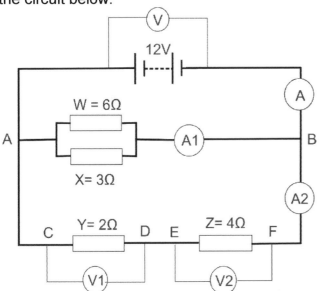

Find the total resistance of the circuit.

Resistance between AB (W and X in parallel)

$1/R = 1/R_1 + 1/R_2$

$\quad = 1/6 + 1/3$

$\quad = ((1x1)+(2x1))/6$

$1/R = 3/6$

$R = 6/3$

$\quad = 2\ \Omega$

Resistance between CF (Y and Z in series)

$R = R_1 + R_2$

$\quad = 2+4$

$\quad = 6\ \Omega$

AB and CF are parallel to each other
So, total resistance in the circuit is

$1/R = 1/R_1 + 1/R_2$

$\quad = 1/2 + 1/6$

$\quad = ((1x3)+(1x1))/6$

$1/R = 4/6$

$R = 6/4$

$\quad = 1.5\ \Omega$

Calculate the current in the ammeter labeled A.

$I = V / R$

$\quad = 12 / 1.5$

$\quad = 8A$

Calculate the current in A1.

$I = V/R$

$\quad = 12 / 2$

$\quad = 6A$

Calculate the current in A2.

Current in AB + current in CF = 8A

$A_1 + 6 = 8$

$\quad = 8 - 6$

$\quad = 2A$

OR

$I = V / R$

$\quad = 12 / 6$

$\quad = 2A$

Calculate the potential difference in the voltmeter V1.

$V = I x R$

$\quad = 2 X 2$

$\quad = 4V$

Calculate the potential difference in the voltmeter V2.

$V = I x R$

$\quad = 2 X 4$

$\quad = 8V$

Calculate the current in W.

$I = V / R$

$\quad = 12 / 6$

$\quad = 2A$

Calculate the current in X.

$I = V / R$

$\quad = 12 / 3$

$\quad = 4A$

The figure of the circuit below shows all the results.

Resistors in parallel and series

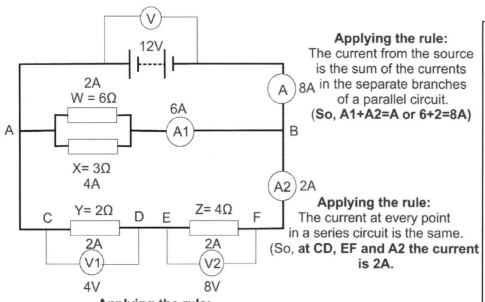

Applying the rule:
The current from the source is the sum of the currents in the separate branches of a parallel circuit.
(So, A1+A2=A or 6+2=8A)

Applying the rule:
The current at every point in a series circuit is the same.
(So, **at CD, EF and A2 the current is 2A.**

Applying the rule:
The sum of the potential difference in a series circuit is equal to the potential difference across the whole circuit. **(The potential difference across V1 and V2 is equal to V=12V)**

Remember:
Circuit rules
- *The current at every point in a series circuit is the same.*
- *The sum of the potential differences in a series circuit is equal to the potential difference across the whole circuit.*
- *The current from the source is the sum of the currents in the separate branches of a parallel circuit.*
- *Use the formula $R = V / I$, to find the current, potential difference or resistance at any point in the circuit.*

Chapter Twenty One
Practical Electricity

21 (a) Electricity

Electricity plays a vital role in our daily lives. Heating of houses in cold countries, lighting of houses, shopping malls and streets and running of motors in water pumps, lifts, etc. are all possible only with the use of electricity. The electricity that is supplied to houses is typically an alternating current (a.c.) with a voltage of 230V and has a frequency of 50 Hz.

21 (b) Power

The power is the rate at which an appliance uses electrical energy. Power is measured in Watts (W) or kilowatts (kW).. it can be calculated by using the following equation.

$$\text{Power} = \text{Voltage} \times \text{Current}$$
$$P = V \times I$$

The power rating of an appliance is important in determining the fuse rating and the gauge of wire that has to be used for the appliance.

Energy

The amount of electrical energy used can be calculated by the product of power and time.

$$\text{Energy} = \text{Voltage} \times \text{Current} \times \text{Time} \qquad \text{or} \qquad \text{Energy} = \text{Power} \times \text{Time}$$
$$E = V \times I \times t \qquad\qquad\qquad E = P \times t$$

The electrical energy used by consumers is calculated in kWh. Electricity companies charge consumers according to the energy consumed.

21 (c) Worked example

A person is buying a microwave oven. The ratings of two ovens in the shop are 800W and 700W. The electricity in the house is supplied by a 240V mains supply.

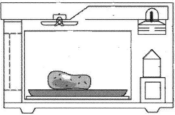

Explain what is meant by a rating of 700W.

The microwave will use 700 Joules of electrical energy every second.

Calculate the current passing through the 800W microwave oven.

$P = V \times I$
$800 = 240 \times I$
$I = 800 / 240$
$= 3.3 A$

Calculate the current passing through the 700W microwave oven.

$P = V \times I$
$700 = 240 \times I$
$I = 700 / 240$
$= 2.9 A$

Calculate the electrical energy consumed by using the 700W microwave oven for 9 minutes.

$E = P \times t$
$= (700 / 1000) \times (9 / 60)$
$= 0.7 \times 0.15$
$= 0.1 \, kWh$

> Remember:
> 1kW = 1000 W
> 1min = 60 s
> 1hr = 3600 s

Calculate the cost of electricity if the microwave is used for 9 minutes per day for 30 days, at the cost of 5 pence per kWh.

$E = P \times t$
$= (700 / 1000) \times (270 / 60)$
$= 0.7 \times 4.5$
$= 3.15 \, kWh$
$Cost = consumption \times charges \, per \, unit$
$= 3.15 \times 5$
$= 15.75 \, pence$

21 (d) Insulation

The electrical conductors are covered with insulators as a safety measure to prevent electrical shock to users and to prevent short-curcuits.

Insulation is made up of non-conducting material like PolyVinylChloride (PVC or plastic). It is an excellent insulator, flexible enough to bend around corners and cheap to make. Older wires had a rubber material as their insulation but the rubber cracked and split as it got older and so it has been replaced in houses by new PVC covered cable. Insulation can become unsafe if it is damaged, as shown in figure 21.1, or if it is wet because impure water will conduct electricity.

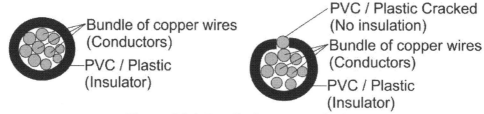

Figure 21.1: Insulation around wires

Overheating of wires due to excessive current may cause the insulation to melt or ignite. This can be prevented by using a fuse or circuit breaker with the appropriate rating.

21 (e) Fuse ratings

A fuse is a safety device which switches off an appliance if too large a current flows through the Live wire. The fuse is connected between the Live pin and the Live wire of a plug. The fuse has a rating printed on the outside in amperes. If the current going through the fuse rises above its rated value, then the fuse melts and turns off the appliance. This helps to protect the appliance from damage by excessive current. For example, if a microwave has a power rating of 700W and 240V then the current flowing through it will be 2.9A. A 3A fuse will be fitted into the microwave. If the current exceeds 3A then the fuse will melt and the microwave will shut down.

21 (f) Earthing

Electrical appliances with metal cases and no Earthing can deliver a fatal shock if the live wire comes in contact with the metal casing, as shown in the figure 21.2. However, if the Earth wire is connected, then most current will flow through the copper wire into the earth and the very little current flows through the person, as a higher current will always flow through a path with lower resistance. The high current flowing through the Earth wire will also heat the wires and cause the fuse to melt or the circuit breaker to trip.

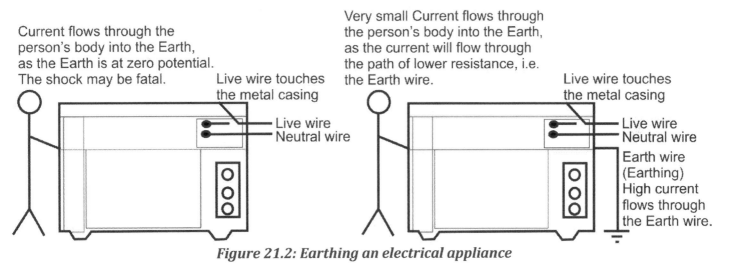

Figure 21.2: Earthing an electrical appliance

21 (f) Double insulation

Electrical appliances, such as vacuum cleaners and electric drills, do not have an earth wire. This is because they have plastic casings and are designed in such a way that the live wire cannot touch the casing. As a result, the casing cannot give an electric shock, even if the wires inside become loose. These appliances have double insulation and carry a symbol, shown in figure 21.3.

Figure 21.3: Symbol for double insulation

21 (g, i) Live, neutral and earth wires

The mains electricity supply has three wires. They are called Live, Neutral and Earth.

The Live wire is connected directly to the generators of the electricity supply company and is maintained at a high potential.

The Neutral wire returns the electricity to the generator after it has passed through the appliance, to complete the circuit. It is maintained at a low potential. When the circuit is complete, current flow from high to low potential.

The Earth wire usually carries no electricity, it is there as a safety device. It is made up of copper wire with high conductivity and very low resistance. The Earth wire is connected to the Earth at zero potential.

The figure 21.4 shows the connection of live, neutral and earth wires in an electrical appliance.

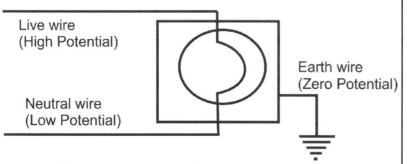

Figure 21.4: Live wire, neutral wire and earth wire

Remember:
Switches, fuses and circuit breakers are always wired into the live conductor. This ensures that the current does not flow through the appliance if the fuse melts, or if the circuit breaker trips, or if the switch is turned off. This protects the appliance from being damaged by high currents. If the fuse or switch was in the neutral wire, then current can still pass from the live wire through the earthing and damage the appliance..

21 (h) Wiring a mains plug safely
The three pin plug has three terminals and are connected, as shown in figure 21.5.

The earth pin is long and larger than the neutral and live pin to ensure that the earth wire makes contact before the live wire in the socket. The live wire is connected to a fuse, which acts as a safety device.

Extreme care must be taken to ensure that the live wire is not connected to the Earth pin. This would charge the body of the electrical appliance and could result in a fatal shock or cause the fuse to melt or the circuit breaker to trip.

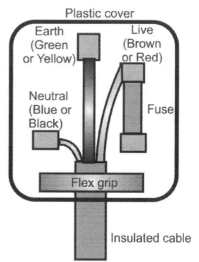

Figure 21.5: Three pin plug

Cambridge 5054 syllabus specification 22 (a) describe experiments to show the force on a current-carrying conductor, and on a beam of charged particles, in a magnetic field, including the effect of reversing (1) the current, (2) the direction of the field.

Cambridge 5054 syllabus specification 22 (b) state the relative directions of force, field and current.

Cambridge 5054 syllabus specification 22 (c) describe the field patterns between currents in parallel conductors and relate these to the forces which exist between the conductors (excluding the Earth's field).

Cambridge 5054 syllabus specification 22 (d) explain how a current-carrying coil in a magnetic field experiences a turning effect and that the effect is increased by increasing (1) the number of turns on the coil, (2) the current.

Cambridge 5054 syllabus specification 22 (e) discuss how this turning effect is used in the action of an electric motor.

Cambridge 5054 syllabus specification 22 (f) describe the action of a split-ring commutator in a two-pole, single-coil motor and the effect of winding the coil onto a soft iron cylinder.

22 (a, b) Force on a current-carrying conductor in a magnetic field

When a current flows through a conductor in a magnetic field, a force is exerted on the conductor. The **current**, **force** and **magnetic field** are **mutually perpendicular to each other**. The direction of each component can be determined by Fleming's **left hand rule**, as shown in Figure 22.1.

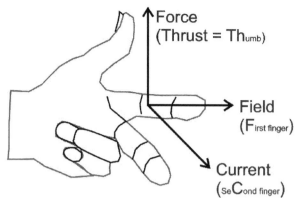

Figure 22.1: Fleming's Left hand rule

The Fleming's Left Hand rule can be demonstrated by setting up an apparatus as shown in figure 22.2 and 22.3. When the switch is closed, a current flows through the conductor and a force is exerted on the conductor. The direction of the force, current and magnetic field is determined by Fleming's Left hand rule.

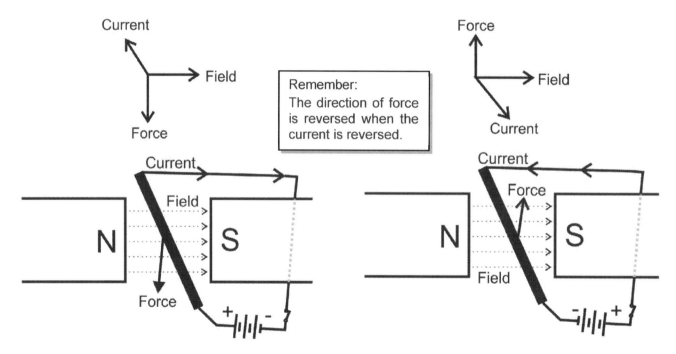

Figure 22.2: Current flows from the positive to negative terminals of the battery, when the switch is closed. Magnetic field is always from N pole towards S pole of the magnets. The force is exerted perpendicular to the magnetic field and current.

Figure 22.3: Current flows from the positive to negative terminals of the battery, when the switch is closed. Magnetic field is always from N pole towards S pole of the magnets. The force is exerted perpendicular to the magnetic field and current.

Force on a beam of charged particles in a magnetic field

A conventional current is the movement of positive charges or protons. So, if protons are beamed through a magnetic field, a force will deflect the proton beam following Fleming's left hand rule, as shown in figure 22.4 and 22.5. The current, magnetic field and force are mutually perpendicular to each other.

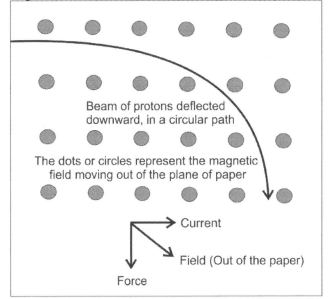

Figure 22.4: The direction of movement of protons is the direction of the conventional current, the magnetic field moves out of the paper and is depicted by dots. The downward force deflects the protons in a circular path.

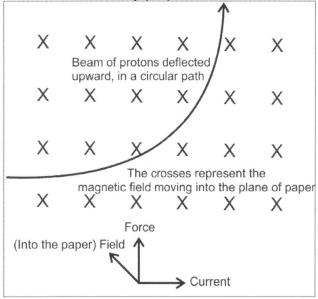

Figure 22.5: The direction of movement of protons is the direction of the conventional current, the magnetic field moves into the paper and is depicted by crosses. The upward force deflects the protons in a circular path.

Note: Changing the direction of the field causes the direction of the force to be reversed.

Likewise, if the direction of current is changed in a magnetic field, the direction of force on the beam of protons is reversed, as shown in figure 22.6 and 22.7.

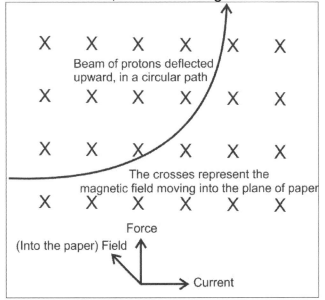

Figure 22.6: The direction of movement of protons is the direction of the conventional current, the magnetic field moves into the paper and is depicted by crosses. The upward force deflects the protons in a circular path.

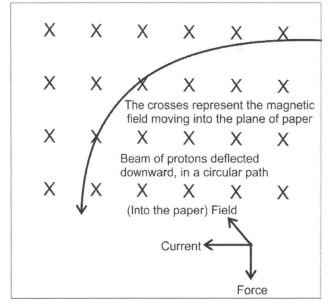

Figure 22.7: The direction of movement of protons is the direction of the conventional current, the magnetic field moves into the paper and is depicted by crosses. The downward force deflects the protons in a circular path.

22 (c) Magnetic Field patterns around a current carrying conductor

When a current is passed through a conductor, a circular magnetic field is generated around the conductor. The direction of the current and the magnetic field is determined by the Right Hand grip rule, as shown in figure 22.8.

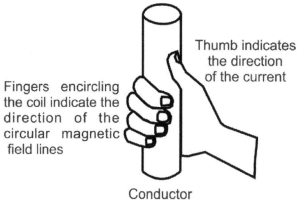

Figure 22.8: Right hand grip rule

When a current carrying conductor is passed through a hole on a cardboard and the magnetic lines are traced by using a compass needle, the field pattern obtained is shown in figure 22.9 and figure 22.10.

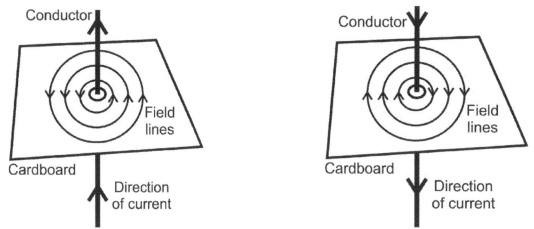

Figure 22.9: Magnetic field lines   *Figure 22.10: Magnetic field lines*

Note: the direction of the field is reversed when the direction of current is reversed.

Forces between current carrying conductors

A magnetic field is generated around a current carrying conductor. If the direction of current in two parallel conductors is the **same**, then the conductors will **attract** each other, as shown in figure 22.11.

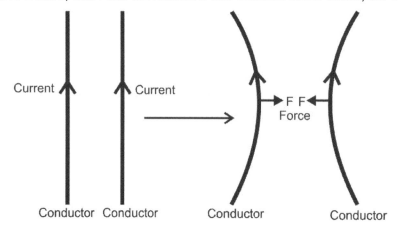

Figure 22.11: Forces between two parallel conductors

The field pattern around each conductor and the resultant field pattern which causes the forces to act on the conductors is shown in figure 22.12.

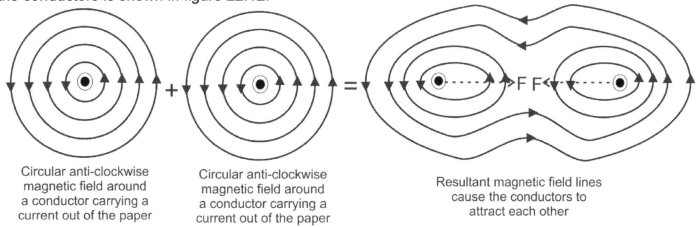

Circular anti-clockwise
magnetic field around
a conductor carrying a
current out of the paper

Circular anti-clockwise
magnetic field around
a conductor carrying a
current out of the paper

Resultant magnetic field lines
cause the conductors to
attract each other

Figure 22.12: Field pattern around two parallel conductors

A magnetic field is generated around a current carrying conductor. If the direction of current in two parallel conductors is the **opposite**, then the conductors will **repel** each other, as shown in figure 22.13.

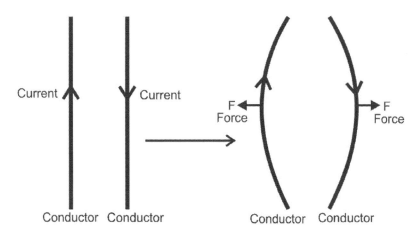

Figure 22.13: Forces between two parallel conductors

The field pattern around each conductor and the resultant field pattern which causes the forces to act on the conductors is shown in figure 22.14.

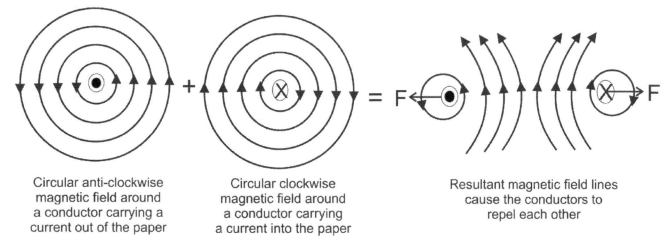

Circular anti-clockwise
magnetic field around
a conductor carrying a
current out of the paper

Circular clockwise
magnetic field around
a conductor carrying
a current into the paper

Resultant magnetic field lines
cause the conductors to
repel each other

Figure 22.14: Forces between two parallel conductors

22 (d) Turning effect of a current on a coil

When a current is passed through a rectangular coil in a magnetic field, the current in either side of the coil flows in opposite directions. This causes the force on the coil to act in opposite directions and parallel to each other. This causes the coil to rotate about its axis. The forces acting on either side of a rectangular coil and the combined magnetic field is shown in figure 22.15.

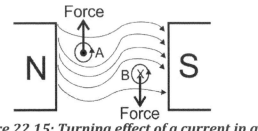

Figure 22.15: Turning effect of a current in a rectangular coil

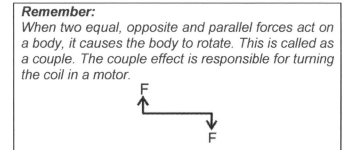

Remember:
When two equal, opposite and parallel forces act on a body, it causes the body to rotate. This is called as a couple. The couple effect is responsible for turning the coil in a motor.

The magnitude of the force exerted on the coil is proportional to the
- the number of turns in the coil, and
- the current passing through the coil.
- Strength of the magnetic field.

22 (e, f) D.C motor
When a current is passed through the coil of the D.C motor, the current in AB flows in an opposite direction to CD. This causes the forces on AB and CD to be equal, opposite and parallel. The coil then begins to rotate. The split rings maintain the contact between the circuit and the coil. The carbon brushes are in contact with the split rings. Winding the coil around a soft iron core increases the magnitude of the force on the coil.

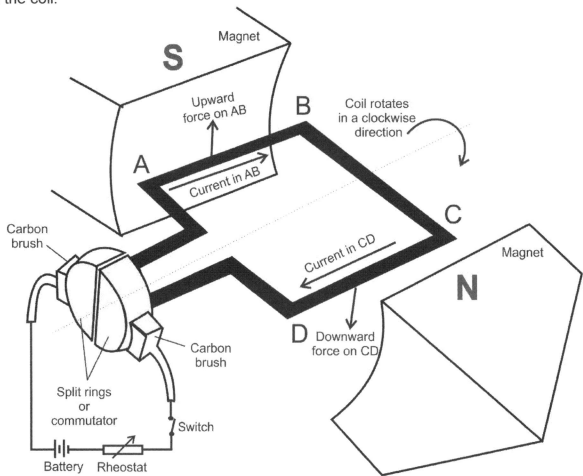

Figure 22.16: Working of a D.C motor

Chapter Twenty Three
Electromagnetic Induction

Cambridge 5054 syllabus specification 23 (a) *describe an experiment which shows that a changing magnetic field can induce an e.m.f. in a circuit.*
Cambridge 5054 syllabus specification 23 (b) *state the factors affecting the magnitude of the induced e.m.f.*
Cambridge 5054 syllabus specification 23 (c) *state that the direction of a current produced by an induced e.m.f. opposes the change producing it (Lenz's Law) and describe how this law may be demonstrated.*
Cambridge 5054 syllabus specification 23 (d) *describe a simple form of a.c. generator (rotating coil or rotating magnet) and the use of slip rings where needed.*
Cambridge 5054 syllabus specification 23 (e) **sketch a graph of voltage output against time for a simple a.c. generator.*
Cambridge 5054 syllabus specification 23 (f) *describe the structure and principle of operation of a simple iron-cored transformer.*
Cambridge 5054 syllabus specification 23 (g) *state the advantages of high voltage transmission.*
Cambridge 5054 syllabus specification 23 (h) *discuss the environmental and cost implications of underground power transmission compared to overhead lines.*

23 (a, b) Experiment to show that a changing magnetic field can induce an e.m.f. in a circuit

When a bar magnet is moved towards a solenoid (long narrow coil), an e.m.f. is generated in the circuit, as shown in figure 23.1. The center-zero galvanometer deflects as long as the magnet is moving. This is because the magnetic field passing through the coil keeps changing and induces an e.m.f. in the circuit. This is called as electromagnetic induction.

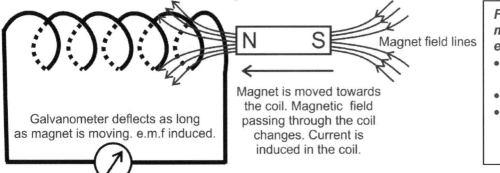

Magnet field lines

Magnet is moved towards the coil. Magnetic field passing through the coil changes. Current is induced in the coil.

Galvanometer deflects as long as magnet is moving. e.m.f induced.

Factors affecting the magnitude of the induced e.m.f.
- *The number of turns in the coil.*
- *The strength of the magnet.*
- *The speed at which the magnet is moved towards or away from the coil.*

Figure 23.1: Electromagnetic induction

If the magnet is kept stationary then no e.m.f. is generated and the galvanometer reading remains at zero. This is because there is no change in the magnetic field passing through the coil, even though the coil may lie within the magnetic field.

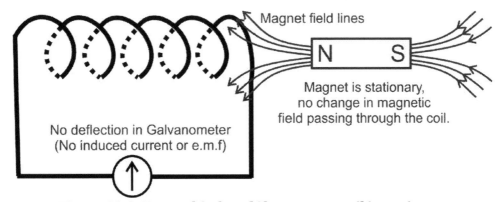

Magnet field lines

Magnet is stationary, no change in magnetic field passing through the coil.

No deflection in Galvanometer (No induced current or e.m.f)

Figure 23.2: No e.m.f. induced if magnet or coil is stationary

The direction of the induced current in the coil can be determined by the following rules, as shown in figure 23.3 and figure 23.4.

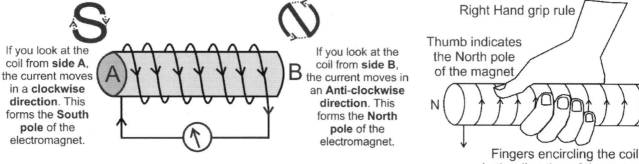

If you look at the coil from side A, the current moves in a **clockwise direction**. This forms the **South pole** of the electromagnet.

If you look at the coil from **side B**, the current moves in an **Anti-clockwise direction**. This forms the **North pole** of the electromagnet.

Right Hand grip rule

Thumb indicates the North pole of the magnet

Fingers encircling the coil in the direction of the current

Figure 23.3: The direction of induced current in the coil by the clockwise and anti-clockwise rule

Figure 23.4: The direction of induced current in the coil by the right hand grip rule

23 (c) Lenz's Law

The direction of a current produced by an induced e.m.f. opposes the change producing it.

In figure 23.5 the North pole of the bar magnet is moved towards the coil. The current induced in the coil will develop a North pole on the right hand side of the coil. So, the coil repels the bar magnet and opposes the movement of the magnet.

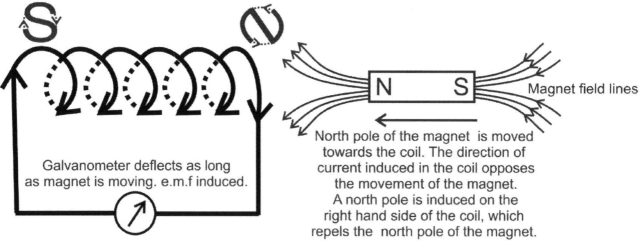

Galvanometer deflects as long as magnet is moving. e.m.f induced.

Magnet field lines

North pole of the magnet is moved towards the coil. The direction of current induced in the coil opposes the movement of the magnet. A north pole is induced on the right hand side of the coil, which repels the north pole of the magnet.

Figure 23.5: Repulsion of North poles oppose the movement of the magnet

In figure 23.6 the North pole of the bar magnet is moved away from the coil. The current induced in the coil will develop a South pole on the right hand side of the coil. So, the coil attracts the bar magnet and opposes the movement of the magnet.

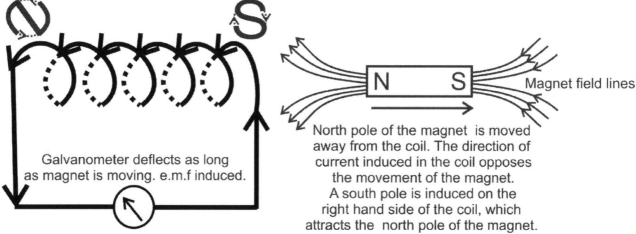

Galvanometer deflects as long as magnet is moving. e.m.f induced.

Magnet field lines

North pole of the magnet is moved away from the coil. The direction of current induced in the coil opposes the movement of the magnet. A south pole is induced on the right hand side of the coil, which attracts the north pole of the magnet.

Figure 23.6: Attraction of North pole and South pole opposes the movement of the magnet

23 (d) Alternating current generator

An alternating current generator works by rotating a rectangular coil in a magnetic field. A current is induced in the coil. As the coil cuts through the magnetic field, the magnetic lines of force passing through the coil changes continuously and an e.m.f. is induced in the coil. The slip rings are connected to the ends of the coil and the carbon brushes conduct the current into the output circuit.

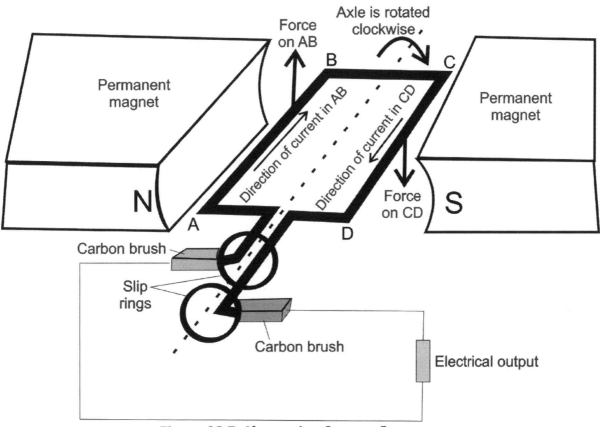

Figure 23.7: Alternating Current Generator

23 (e) Voltage output for a simple a.c. generator

As the coil is rotated in the magnetic field, the direction of current in the coil gets reversed and the magnitude of the induced e.m.f. changes depending on the position of the coil as shown in figure 23.8.

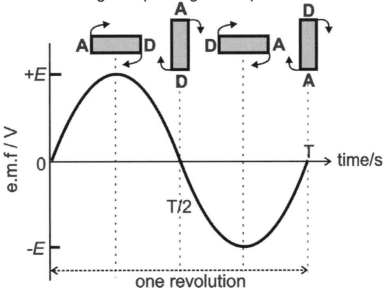

Figure 23.8: Changes in e.m.f. against time

The magnitude of the induced e.m.f. depends upon
- the number of turns in the coil, and
- the speed at which the coil is rotated.
- the strength of the magnetic field.

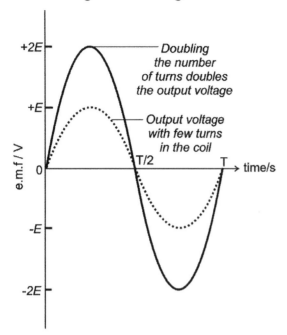

Figure 23.9: Increasing the number of turns increases the e.m.f induced in the coil.

Figure 23.10: Increasing the speed of the coil increases the e.m.f induced in the coil.

23 (f) Working principle of a Transformer

A transformer works on the principle of **mutual induction**. An alternating current in a primary coil generates and induced e.m.f. in the secondary coil. As shown in figure 23.8, an alternating current keeps changing its magnitude and direction with time. This will cause the magnetic field of the primary coil to change continuously and cut through the secondary coil. The continuous change in the magnetic field through the secondary coil generates an induced e.m.f in the secondary coil. The frequency of the induced current in the secondary coil is the same as the frequency of the alternating current in the primary coil. The magnitude of the current in the secondary coil depends upon the relative number of turns in the primary and secondary coil.

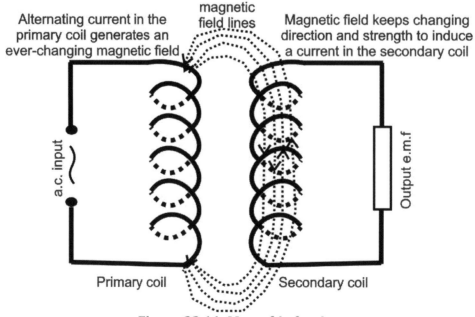

Figure 23.11: Mutual induction

Step up transformer

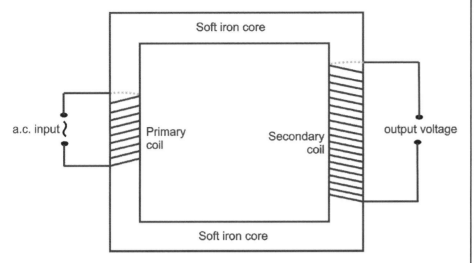

Figure 23.12: Step up transformer

Remember:
In a step up transformer, the number of turns in the secondary coil (N_s) is more than the number of turns in the primary coil (N_p).

$$N_S > N_p$$

Since, the secondary coil has more number of turns than the primary coil, the output voltage (V_s) is larger than the input voltage (V_p).

$$V_S > V_p$$

The equation relating the output voltage, input voltage and number of turns in the coils is given below.

$$V_S/V_p = N_S/N_p$$

Step down transformer

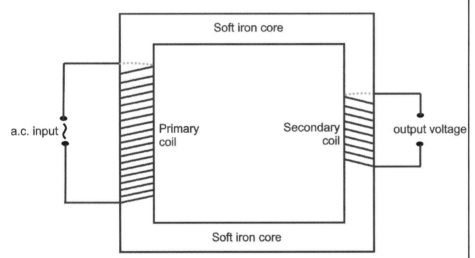

Figure 23.13: Step down transformer

Remember:
In a step down transformer, the number of turns in the secondary coil (N_s) is less than the number of turns in the primary coil (N_p).

$$N_S < N_p$$

Since, the secondary coil has less number of turns than the primary coil, the output voltage (V_s) is smaller than the input voltage (V_p).

$$V_S < V_p$$

The equation relating the output voltage, input voltage and number of turns in the coils is given below.

$$V_S/V_p = N_S/N_p$$

The soft iron core encloses the magnetic field and allows all the magnetic lines from the primary coil to pass through the secondary coil. The soft iron core is made of many thin laminated plates to reduce energy losses due to local currents (Eddy currents) in the soft iron core itself.

Power transfer in a transformer

Assuming that the efficiency of the transformer is ideal (100%),

Power input in the primary coil = Power output in secondary coil
$$V_p \, I_p = V_s \, I_s$$

Combining all the two equations we obtain the following equation

$$V_s / V_p = N_s / N_p = I_p / I_s$$

Where,
I_p is the current in the primary coil and I_s is the current in the secondary coil.

Remember:

Inducing a high voltage in the secondary coil, reduces the current in the secondary coil.

Likewise, reducing the voltage increases the current.

23 (g) Advantages of high voltage transmission

Power is transmitted from the powerhouse to industries and households through copper or aluminum cables. Energy is lost as heat during transmission of current.

A step up transformer is used to raise the voltage from 25,000 V to 400,000 V. This reduces the current from 8,000 A to 500 A.

The advantages of high voltage and low current transmission are:
- less energy is lost during transmission
- cables of lower rating can be used for transmission. So, the cost of cables is reduced.

23 (h) Underground power transmission compared to overhead lines

Copper cables carrying the electricity are buried in the ground or aluminum cables are suspended from pylons.

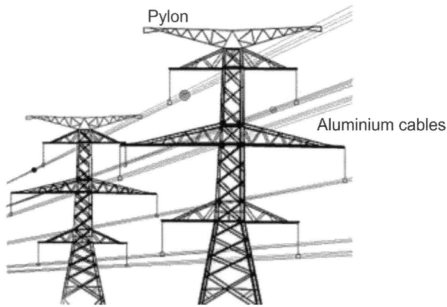

Figure 23.14: Power transmission through aluminium cables

The advantages and disadvantages are listed below.
- Aluminium cables have a low density and can safely be suspended from inexpensive thin pylons, as shown in figure 23.14. The pylons also make it easy to access the cables for maintenance and repair.
- However, pylons are disadvantageous as they look ugly on the landscape.
- Electricity Transmission using pylons is cheaper than burying and repairing copper cables underground.

Chapter Twenty Four
Introductory Electronics

Cambridge 5054 syllabus specification 24 (a) state that electrons are emitted by a hot metal filament.

Cambridge 5054 syllabus specification 24 (b) explain that to cause a continuous flow of emitted electrons requires (1) high positive potential and (2) very low gas pressure.

Cambridge 5054 syllabus specification 24 (c) describe the deflection of an electron beam by electric fields and magnetic fields.

Cambridge 5054 syllabus specification 24 (d) state that the flow of electrons (electron current) is from negative to positive and is in the opposite direction to conventional current.

Cambridge 5054 syllabus specification 24 (e) describe in outline the basic structure and action of a cathode-ray oscilloscope (c.r.o.) (detailed circuits are not required).

Cambridge 5054 syllabus specification 24 (f) describe the use of a cathode-ray oscilloscope to display waveforms and to measure p.d.s and short intervals of time (detailed circuits are not required).

Cambridge 5054 syllabus specification 24 (g) explain how the values of resistors are chosen according to a colour code and why widely different values are needed in different types of circuit.

Cambridge 5054 syllabus specification 24 (h) discuss the need to choose components with suitable power ratings.

Cambridge 5054 syllabus specification 24 (i) describe the action of thermistors and light-dependent resistors and explain their use as input sensors (thermistors will be assumed to be of the negative temperature coefficient type).

Cambridge 5054 syllabus specification 24 (j) describe the action of a variable potential divider (potentiometer).

Cambridge 5054 syllabus specification 24 (k) describe the action of a diode in passing current in one direction only.

Cambridge 5054 syllabus specification 24 (l) describe the action of a light-emitting diode in passing current in one direction only and emitting light.

Cambridge 5054 syllabus specification 24 (m) describe the action of a capacitor as a charge store and explain its use in time-delay circuits.

Cambridge 5054 syllabus specification 24 (n) describe and explain the action of relays in switching circuits.

Cambridge 5054 syllabus specification 24 (o) describe and explain circuits operating as light-sensitive switches and temperature-operated alarms (using a relay or other circuits).

24 (a, b) Thermionic emission

Thermionic emission is the release of electrons from heated metal filaments, as shown in figure 24.1. Tungsten is commonly used as the filament as it has a high melting point.

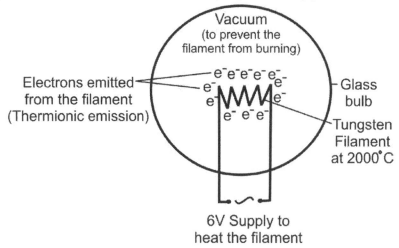

Figure 24.1: Thermionic emission

The electrons which are emitted by the heated filament can be attracted towards an anode at a high potential, as shown in figure 24.2.. The low gas pressure reduces collisions between the electrons and gases and prevents the filament from burning. The electrons are accelerated towards the anode. This is commonly used in cathode ray tubes.

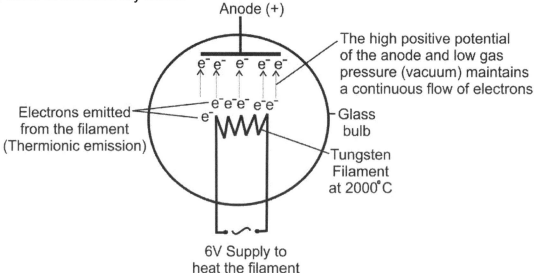

Figure 24.2: Electrons flow towards the anode

24 (d) Remember:

- *Electron beams are deflected by electric fields and magnetic fields.*
- *The flow of electrons (electron current) is from negative to positive and is in the opposite direction to conventional current.*
- *Refer to figures 22.4, 22.5, 22.6 and 22.7 to determine the deflection of electron beams and current.*

24 (c, e) Structure and action of a cathode-ray oscilloscope (c.r.o.)

The cathode-ray oscilloscope (C.R.O.) consists of the following components:

1. The electron gun

Parts of Electron Gun	Function
Filament	To heat the cathode.
Cathode	Release electrons when heated by filament.
Grid	The grid is connected to a negative potential. The more negative this potential, the more electrons will be repelled from the grid and fewer electrons will reach the anode and the screen. The number of electrons reaching the screen determines the brightness of the light. Hence, the negative potential of the grid can be used as a brightness control.
Focusing Anode and Accelerating anode	The other feature in the electron gun is the use of the anode. The anode at positive potential accelerates the electrons and the electrons are focused into a fine beam as they pass through the anode.

2. The deflecting plates

Part of the deflecting system	Function
Y-plate	The Y-plates will cause deflection in the vertical direction when a voltage is applied across them. It is called the gain control.
X-plate	On the other hand, the X-plates will cause the electron beam to be deflected in the horizontal direction if a voltage is applied across them. It is called the time base.

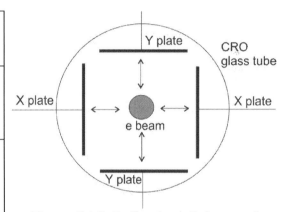

Figure 24.3: Deflection of electron beam

3. The fluorescent screen

The screen is coated with a fluorescent salt, for example, zinc sulphide. When the electrons hit the screen, it will cause the salt to produce a flash of light and hence a bright spot on the screen.

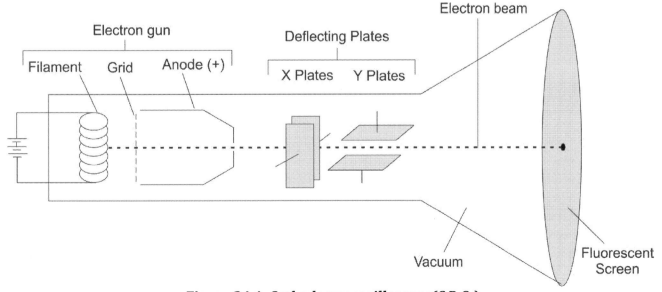

Figure 24.4: Cathode ray oscilloscope(C.R.O.)

24 (f) Figure 24.5 shows a wave obtained with the time-base set to 2ms/cm and the vertical gain at 5V/cm.

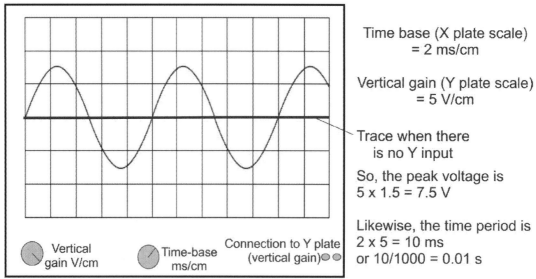

Time base (X plate scale)
= 2 ms/cm

Vertical gain (Y plate scale)
= 5 V/cm

Trace when there is no Y input

So, the peak voltage is
5 x 1.5 = 7.5 V

Likewise, the time period is
2 x 5 = 10 ms
or 10/1000 = 0.01 s

Figure 24.5: Trace of a C.R.O

24 (g) Colour coding of resistors

Resistors of different values are used in electronic circuits. The values of the resistance is marked as coloured bands on the resistors. The colour codes and the rules are shown in figure 24.6. The values of resistors are chosen according to a colour code.

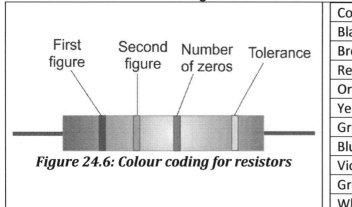

Figure 24.6: Colour coding for resistors

Colour	Number
Black	0
Brown	1
Red	2
Orange	3
Yellow	4
Green	5
Blue	6
Violet	7
Grey	8
White	9

Colour	Gold	Silver	No colour
Tolerance	5%	10%	20%

24 (h) Worked example

A resistor has the colour code as shown in the diagram below. Calculate the resistance of the resistor.

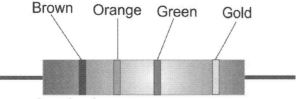

Brown   Orange   Green   Gold

First band is brown = 1
Second band is orange = 3
Third band is green = 00000
So, the resistance = 1300000Ω
Or, 1300kΩ
Or, 1.3MΩ

Calculate the maximum resistance that the resistor may provide.

Since, the tolerance of the resistor is depicted by a gold band, the maximum tolerance is 5% higher than 1300000 Ω.

5% of 1300000 is
(5/100) x 1300000
= 65000Ω

So, maximum resistance = 1300000+65000
= 1,365,000Ω

In electrical and electronic appliances, the power rating of a device is a guideline set by the manufacturer as a maximum power to be used with that device. This limit is usually set somewhat lower than the level where the device will be damaged, to allow a margin for safety.

24 (j) Variable potential divider (potentiometer)

A potential divider is used to divide the voltage from the source into different components in a circuit. Each component can be supplied with a specific voltage by using a resistor of the appropriate resistance. For example if there is a bulb of rating 5V and a battery of 10V, the filament of the bulb will melt if used at a potential of 10V, as shown in figure 24.7. However, if a resistor with a resistance of the same value as the bulb is connected in series with the bulb, as shown in figure 24.8, then the **potential gets divided** and the bulb receives only 5V and does not fuse.

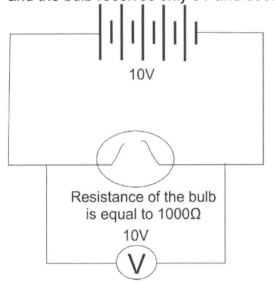

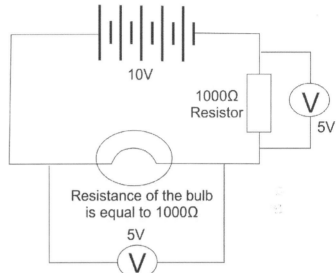

Figure 24.7: Bulb fuses due to high potential   *Figure 24.8: Dividing the potential protects the bulb*

The potential divider used above is a fixed divider. However, if a single variable resistor(R) is included in series with the circuit, then the potential can be divided to any value by sliding S along the length of the resistor AB, as shown in figure 24.9. This form of the potential divider is called a potentiometer.

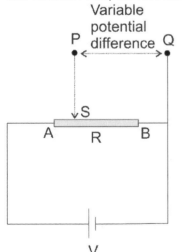

Figure 24.9: Potential divider

> **Remember:**
> The Voltage (V) given out by the source is divided between the resistors in series. So, If there are two appliances then with a potential of V1 and V2 then the sum of V1 and V2 must be equal to V.
>
> Potential dividers and potentiometers are widely used in switches for ceiling lights, volume control for music players and heating control for electric cookers.

24 (m) Capacitor

A capacitor is a device that can store electrical charges. It has many uses in electronics. One of these is as a time delay switch. One of the simplest ways to make a capacitor is to have two metal plates close together, as shown in figure 24.10.

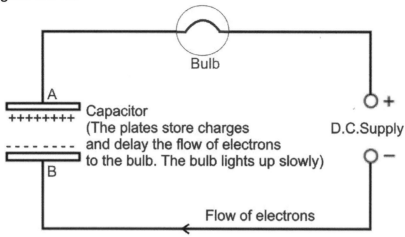

Figure 24.10: Capacitor as a time delay switch

When the circuit is closed, electrical charges start following into the capacitor. The plate in the capacitor that is connected to the positive pole of the battery will start filling up with positive charges.
The other plate connected to the negative pole will fill up with negative charges. This can happen for two reasons:
1. the large surface areas of the plates give the charges more space to accumulate, and
   2. the small distance between the plates means that the opposite charges can attract each other and help keep the charges on the plates.
2. As charges fill up in the capacitor, the voltage across the capacitor increases. Point A gets more and more positive, while point B gets more and more negative. This means that we can actually use A and B as if they are the positive and negative poles of a battery. So if we connect A and B to a light bulb, the bulb will light up.

But the bulb will not light up immediately. It will start dim and grow brighter until it reaches the full brightness. This is because the capacitor takes time to charge up. It is fully charged when the voltage across A and B gets as big as the voltage of the battery, because then the battery's voltage will not be enough to force more charges into the capacitor. The time taken for the capacitor voltage to reach its maximum depends on:
1. the size of the capacitor, and
2. the value of the resistance in the circuit.

The larger these are, the longer it takes for the capacitor to be fully charged. A light bulb connected to it will also take longer to fully light up. In this way, the capacitor can be used as a time delay switch.

24 (n, o, i) Reed relays
The reed switch is made up of two strips of soft iron, almost touching each other, enclosed in a glass case. The strips of soft iron can be moved by a magnetic field to either switch on the circuit or switch it off. Some reed switches are normally open (NO) and can be switched off by magnetic fields and others are normally closed (NC) and can be switched on by magnetic fields. The terminals A and B are used to connect the switch to the circuit.

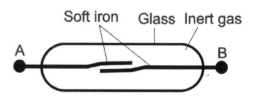

Figure 24.11: A reed switch

The figures 24.12 and 24.13 show the LDR and reed relay working together to control a light sensitive alarm.

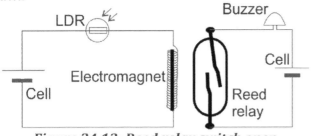 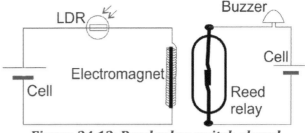

Figure 24.12: Reed relay switch open   *Figure 24.13: Reed relay switch closed*

- When the light intensity is low, the resistance of the circuit is high.

- So, the current in the circuit is low and the electromagnet strength is very low.

- The reed switch remains open and the buzzer does NOT ring.

- When the light intensity increases, the resistance of the circuit decreases due to the LDR.
- So, the current in the circuit increases and the electromagnet strength increases.
- The electromagnet induces magnetic poles in the soft iron strips of the reed switch and the switch is closed. The buzzer rings and sets the alarm on.

The figures 24.14 and 24.15 show a thermistor and reed relay working together to control a temperature sensitive fire alarm.

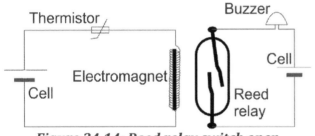 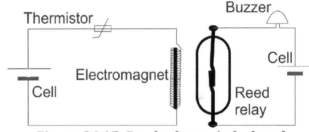

Figure 24.14: Reed relay switch open   *Figure 24.15: Reed relay switch closed*

- When the Temperature is low, the resistance of the circuit is high.

- So, the current in the circuit is low and the electromagnet strength is very low.

- The reed switch remains open and the buzzer does NOT ring.

- When the Temperature increases, the resistance of the circuit decreases due to the thermistor
- So, the current in the circuit increases and the electromagnet strength increases.
- The electromagnet induces magnetic poles in the soft iron strips of the reed switch and the switch is closed. The buzzer rings and sets the alarm on.

24 (k, l) Diodes and Light Emitting Diodes (LEDs)
Diodes and LEDs are used to allow current to flow in one direction only. The LEDs also emit light. LEDs are very small and cheap. They are commonly used in electronic circuits for illumination of displays. The LEDs save energy and reduce the cost of electric consumption.

Cambridge 5054 syllabus specification 25 (a) describe the detection of alpha-particles, beta-particles and gamma-rays by appropriate methods.

Cambridge 5054 syllabus specification 25 (b) state and explain the random emission of radioactivity in direction and time.

Cambridge 5054 syllabus specification 25 (c) state, for radioactive emissions, their nature, relative ionising effects and relative penetrating powers.

Cambridge 5054 syllabus specification 25 (d) describe the deflection of radioactive emissions in electric fields and magnetic fields.

Cambridge 5054 syllabus specification 25 (e) explain what is meant by radioactive decay.

Cambridge 5054 syllabus specification 25 (f) explain the processes of fusion and fission.

Cambridge 5054 syllabus specification 25 (g) describe, with the aid of a block diagram, one type of fission reactor for use in a power station.

Cambridge 5054 syllabus specification 25 (h) discuss theories of star formation and their energy production by fusion.

Cambridge 5054 syllabus specification 25 (i) explain what is meant by the term half-life.

Cambridge 5054 syllabus specification 25 (j) make calculations based on half-life which might involve information in tables or shown by decay curves.

Cambridge 5054 syllabus specification 25 (k) describe how radioactive materials are moved, used and stored in a safe way.

Cambridge 5054 syllabus specification 25 (l) discuss the way in which the type of radiation emitted and the half-life determine the use for the material.

Cambridge 5054 syllabus specification 25 (m) discuss the origins and effect of background radiation.

Cambridge 5054 syllabus specification 25 (n) discuss the dating of objects by the use of $_{14}$C.

25 (a) Detection of radioactivity

Radioactive substances spontaneously release three types of radiation,

Alpha (α) particles
Beta (β) particles
Gamma (γ) rays

The table summarises some of the properties of radiation.

	Charge	Relative Mass
Alpha	+ 2	4
Beta	- 1	Very small
Gamma	0	0

Detection of radioactivity
Geiger-Müller tube

Radiation enters the GM tube through the mica window. The radioactivity ionizes the argon gas in the tube. This causes the charged argon gas to flow between the anode and cathode. The collision of electrons with the positive electrode generates a tiny amount of electricity, which causes **a click** from a loudspeaker of the counter. The number of clicks is a measure of the radioactivity being detected by the tube.

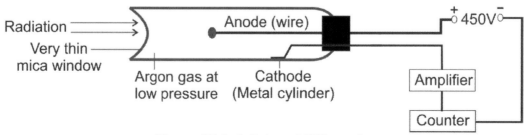

Figure 25.1: A Geiger-Müller tube

Gold leaf electroscope

Dry air is normally a good insulator, so a charged electroscope will stay that way, as the charge cannot escape.

When an electroscope is charged, the gold leaf sticks out, because the charges on the gold repel the charges on the metal stem.

When a **radioactive source comes near**, the air is ionised, and starts to conduct electricity. This means that the charge can "leak" away, the electroscope discharges and **the gold leaf falls**.

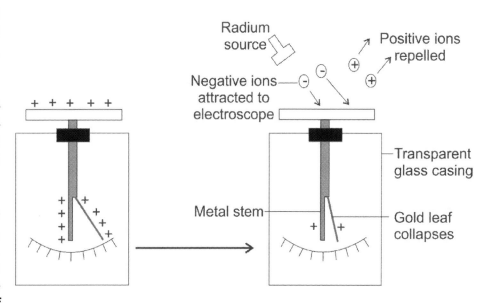

Figure 25.2: A gold leaf electroscope

Cloud chamber

The chamber contains a supersaturated vapour of methylated spirits. Radiation causes ionization of the air. Alcohol and water vapour condenses on the ions produced and form tracks.

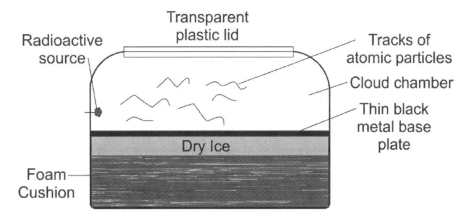

Figure 25.3: A cloud chamber

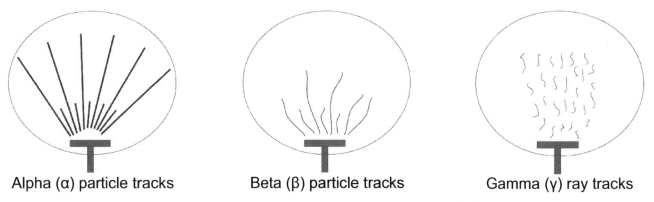

Alpha (α) particle tracks   Beta (β) particle tracks   Gamma (γ) ray tracks

Figure 25.4: Tracks formed by radiation in a cloud chamber

Photographic film badge

A film badge is a thin plastic container which opens at a hinge. Inside there is a piece of film behind some windows. People who work with radioactivity (or X-rays) wear a film badge to monitor their exposure. Radiation will make photographic film darken in the same way that exposure to light and X-rays do. The greater the amount of radiation that the film is exposed to, the darker the film becomes, as shown in figure 25.5.

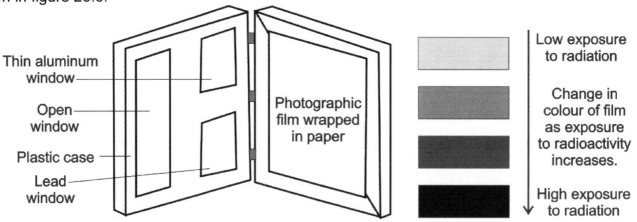

Figure 25.5: Radiation exposure badge

25 (b) Random emission of radioactivity

Radioactive nuclei are rather unstable. The nuclei decay and release Alpha (α) particles, Beta (β) particles or Gamma (γ) rays to make the nuclei more stable.

Radioactivity is spontaneous which means that it is not influenced by any physical factors such as temperature, pressure, time, etc.

The emission of radioactivity is random. This means that the emission occurs at irregular time intervals and the type of radiation given out is also random. This is because when an α or a β particle is released by decay, there is a change in the energy associated with the nucleus. This energy is released as γ radiation.

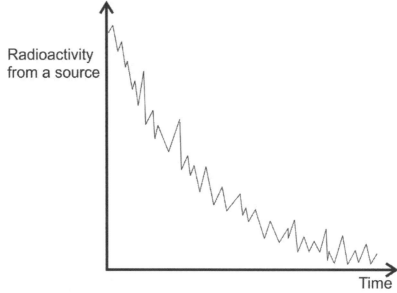

Figure 25.6: Random emission of radioactivity

25 (c) Radioactive emissions: nature, relative ionising effects and relative penetrating powers

Characteristic	Alpha particle (α)	Beta particle (β)	Gamma ray (γ)
Nature	Positively charged helium nucleus	Negatively charged electron	Neutral electromagnetic radiation
In an electric field	Bends to the negative plate	Bends to the positive plate	Does not bend, showing that it is neutral.
In magnetic field	Bends a little showing that it has a big mass. Direction of the bend indicates that it is positively charged.	Bends a lot showing that it has a small mass. Direction of the bend indicates that it is negatively charged.	Does not bend showing that it is neutral.
Ionising power	Strongest	Intermediate	Weakest
Penetrating power	low	Intermediate	High
Stopped by	A thin sheet of paper	A few millimeters of aluminium	A few centimeters of lead or concrete
Range in air	A few centimeters	A few metres	A few hundred metres

The figure 25.7 illustrates an experiment to determine the penetrating power of the three types of radioactivity. A GM tube connected to a count meter is used to detect the count rate for a source. The results obtained are as follows:

- The count rate for alpha particles decreases as a sheet of paper is placed between the source and the GM tube. This indicates that alpha particles are stopped by a sheet of paper.
- The count rate for beta particles shows no decrease with a sheet of paper between the source and the GM tube. However, the count rate decreases when an aluminum foil of 5 mm thickness is placed between the source and the GM tube. This indicates that beta particles are stopped by an aluminum sheet.
- The count rate for gamma shows no decrease with a sheet of paper and an aluminum sheet placed between the source and the GM tube. However, the count rate decreases when a block of lead metal of 2 cm thickness is placed between the source and the GM tube. This indicates that gamma radiation can only be stopped by a thick block of lead.

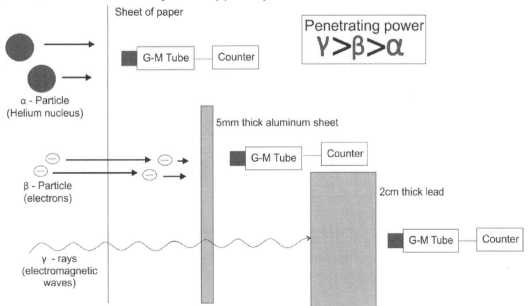

Figure 25.7: Penetration of radioactivity

25 (d) Deflection of radioactive emissions in electric fields

Alpha and beta particles are charged and deflected by an electric field, as shown in figure 25.8. Gamma rays are not charged and are not deflected by electric fields.

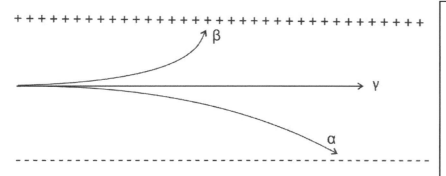

Remember:
- Alpha particles bend towards the negative plate.
- Beta particles bend towards the positive plate.
- Gamma rays remain undeflected, as they are not charged.
- Alpha deflection is less, as it is heavy.
- Beta deflection is more, as it is lighter.

Figure 25.8: Deflection of radioactive emission by electric fields

Deflection of radioactive emissions in magnetic fields.

Alpha and beta particles are charged and deflected by a magnetic fields, as shown in figure 25.9. Gamma rays are not charged and are not deflected by electric fields.

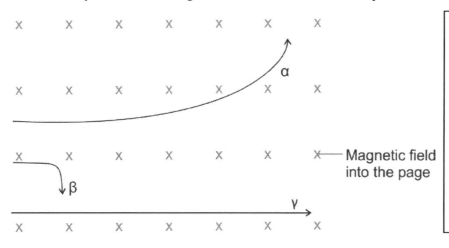

Remember:
- Gamma rays remain undeflected, as they are not charged.
- Alpha deflection is less, as it is heavy.
- Beta deflection is more, as it is lighter.
- Refer to figure 22.1 to determine the direction of deflection.
- Current direction is the same as direction of alpha particles.
- The direction of beta particles is opposite to the direction of current.

Figure 25.9: Deflection of radioactive emission by magnetic fields

25 (e) Radioactive decay

Radioactive nuclei will give out alpha particle, beta particles and gamma radiation. The release of alpha and beta particles causes a change in the composition of the nuclide. This is called radioactive decay.

The changes in the nuclide when an alpha particle is released is shown below.

$$^A_Z X \rightarrow {}^{A-4}_{Z-2} X + {}^4_2 \alpha$$

Example,

$$^{241}_{94} Pu \longrightarrow {}^4_2 \alpha + {}^{237}_{92} U + Energy$$

The changes in the nuclide when a beta particle is released is shown below.

$$^A_Z X \longrightarrow {}^A_{Z+1} X + {}^0_{-1} \beta + Energy$$

Example,

$$^{237}_{92} U \longrightarrow {}^{237}_{93} Np + {}^0_{-1} \beta + Energy$$

When a nuclide releases alpha and beta particles, the nuclide is in the excited state. The excess energy in the nuclide is released as gamma radiation.

Radioactivity is measured in Becquerels, symbol Bq. 1 Bq means that there is one decay per second.

25 (f, g, h) Nuclear fission and nuclear fusion

Nuclear fission is the process in which a large heavy nucleus is split into two smaller nuclei. There is a very tiny change in mass of the daughter nuclei, when compared to the parent nucleus. This is called as the mass defect. The energy released from the reaction is given by the equation

$$E = mc^2,$$

where m is mass defect and c is the velocity of light.

So, nuclear reactions release large amount of energy.

Example of nuclear fission,

$$^{239}_{92}U \longrightarrow \ ^{95}_{38}Sr + \ ^{141}_{54}Xe + 3 \text{ neutrons}$$

Nuclear fusion is the process in which two small nuclei fuse to form a larger heavy nucleus. There is a very tiny change in mass of the daughter nucleus, when compared to the parent nuclei. This is called as the mass defect. The energy released from the reaction is given by the equation

$$E = mc^2,$$

where m is mass defect and c is the velocity of light.

So, nuclear reactions release large amount of energy.

Example of nuclear fusion,

$$^{2}_{1}H + \ ^{2}_{1}H \longrightarrow \ ^{4}_{2}He$$

This reaction occurs in the sun and other stars

25 (i, j) *Half-life*

Half-life is the time taken for half of the radioactive nuclei to decay.

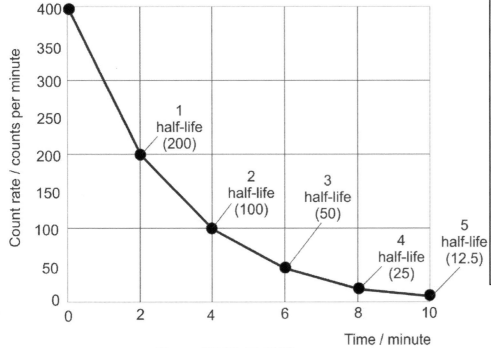

Figure 25.10: Half-life

25 (j) Half-life

The table below and graph alongside show that the count decreases by 50% every 2 minutes. So, the half-life of the source is 2minutes.

Time/min	Count rate/cpm
0	400
2	200
4	100
6	50
8	25
10	12.5

25 (k) Storage and handling of radioactive materials

Since radiation can cause mutation, cancers and cell damage, care must be taken in handling and storing radioactive material. Some of the safety precautions are:

- keep distance from the source by using long metal tongs
- Always point the radioactive source away from the body
- Handle radioactive material only when using a lead apron
- Use radioactive material only for a short time and monitor exposure with a film badge
- Store and transport radioactive material in a lead container, as shown in figure 25.11

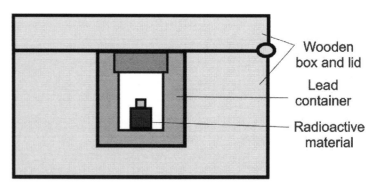

Figure 25.11: Box for storing and transporting radioactive material

25 (l, m, n) Origins and effect of background radiation

Background radiation is naturally occurring, unavoidable and is always present in the surrounding. Background radiation comes from rocks, space, stars, etc.

The background radiation must first be measured and then subtracted from the reading obtained from a source, to get accurate readings.

Radioactive carbon dating

All living things take in carbon from the environment. Plants take in carbon during photosynthesis. Animals take in carbon when they eat food because food contains carbon. All living things therefore have carbon-14 in them at the same amount which is present in the environment. This amount is small. Only one in 850 billion carbon atoms are the isotope carbon-14. The others are not radioactive. When a living thing dies, it stops taking in carbon from its environment. The amount of carbon-14 in it will start to decrease as the carbon-14 slowly decays. The further back in time that something died, the less carbon-14 will be present in it today. The half-life of carbon-14 is approximately 5,700 years. Measuring the amount of carbon-14 in a sample today can tell you how long ago the thing died and therefore the age of the sample, as shown in figure 25.12.

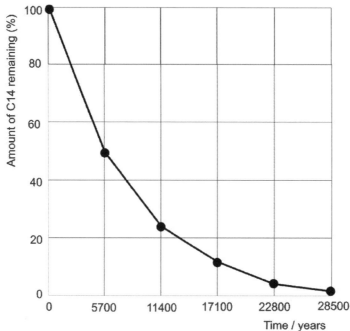

Figure 25.12: Calibration curve for C_{14} half-life

Chapter Twenty Six
The Nuclear Atom

Cambridge 5054 syllabus specification 26 (a) *describe the structure of the atom in terms of nucleus and electrons.*
Cambridge 5054 syllabus specification 26 (b) *describe how the Geiger-Marsden alpha-particle scattering experiment provides evidence for the nuclear atom.*
Cambridge 5054 syllabus specification 26 (c) *describe the composition of the nucleus in terms of protons and neutrons.*
Cambridge 5054 syllabus specification 26 (d) *define the terms proton number (atomic number), Z and nucleon number (mass number), A.*
Cambridge 5054 syllabus specification 26 (e) *explain the term nuclide and use the nuclide notation $^{A}_{Z}X$ to construct equations where radioactive decay leads to changes in the composition of the nucleus.*
Cambridge 5054 syllabus specification 26 (f) *define the term isotope.*
Cambridge 5054 syllabus specification 26 (g) *explain, using nuclide notation, how one element may have a number of isotopes.*

26 (a, c, d) Structure of the atom

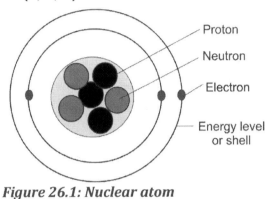

Proton
Neutron
Electron
Energy level or shell

- The nucleus of an atom consists of protons and neutrons.
- The electrons revolve around the nucleus in specific energy levels, called shells.
- The number of protons in the nucleus is called the atomic number (Z).
- The number of protons and neutrons together is called the mass number or nucleon number (A).

Figure 26.1: Nuclear atom

26 (b) Geiger-Marsden alpha-particle scattering experiment

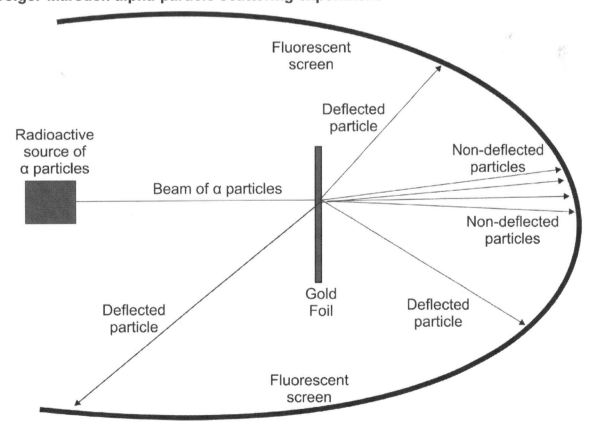

Fluorescent screen

Deflected particle

Radioactive source of α particles

Beam of α particles

Non-deflected particles

Non-deflected particles

Deflected particle

Gold Foil

Deflected particle

Fluorescent screen

Figure 26.2: Geiger-Marsden experiment

Geiger and Marsden designed an experiment to determine the structure of an atom. A beam of alpha particles was aimed at very thin gold foil and their passage through the foil detected. The scientists expected the alpha particles to pass straight through the foil, but something else also happened.

Some of the alpha particles emerged from the foil at different angles, and some even came straight back. The scientists realised that the positively charged alpha particles were being repelled and deflected by a tiny concentration of positive charge in the atom. As a result of this experiment, the nuclear model of the atom was established. It confirmed that the nucleus of the atom is heavy and positively charged. Most spaces around the nucleus are empty spaces and contain the electrons, which have a negligible mass.

26 (e) Nuclides

A **nuclide** is an atomic species characterized by the specific constitution of its nucleus, i.e., by its number of protons, its number of neutrons and its nuclear energy state. The nuclide focuses on the composition of the nucleus.

For example, if X represents the nucleus of an element and A is the nucleon number and Z is the proton number, then the nuclide can be represented symbolically as $^{A}_{Z}X$.

Thus a helium nucleus with 2 protons and 2 neutrons can be represented as $^{4}_{2}He$.

26 (f, g) Isotope

Isotopes are atoms of the same element which have the same proton number but different mass number or neutron number.

For example, Hydrogen has three isotopes, namely Hydrogen, Deuterium and Tritium. The composition and nuclides are represented in the diagrams below.

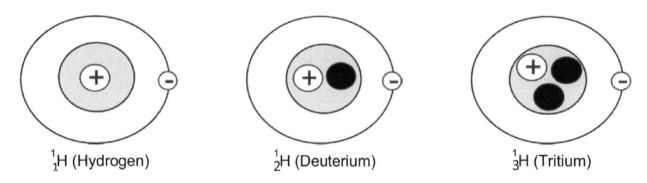

$^{1}_{1}H$ (Hydrogen) $^{1}_{2}H$ (Deuterium) $^{1}_{3}H$ (Tritium)

Figure 26.3: Isotopes of Hydrogen

Visit: www.staffordeducationalservices.com for audiovisual lessons and past paper discussions.

Formula sheet

Chapter 1	
Conversion of units. General rule: • **Larger to smaller units: Multiply** by converting factor. • **Smaller to larger units: Divide** by converting factor. **Units of mass** 1 Kg = 1000g (Converting factor is 1000) 1g = 1000mg (Converting factor is 1000) **Units of length** 1 Km = 1000m (Converting factor is 1000) 1m = 100 cm (Converting factor is 100) 1 cm = 10 mm (Converting factor is 10) 1m = 1000mm (Converting factor is 1000)	**Units of time** 1 hr = 60 mins (Converting factor is 60) 1 min = 60s (Converting factor is 60) 1 hr = 3600s (Converting factor is 3600) **Units of area** $1km^2 = 10^6 \ m^2$(Converting factor is 10^6) $1m^2 = 10^4 \ cm^2$ (Converting factor is 10^4) **Units of volume** $1m^3 = 10^6 \ cm^3$ (Converting factor is 10^6) $1dm^3 = 1000cm^3$ (Converting factor is 1000) **Units of density** $1Kg/m^3 = 1000 \ g/cm^3$ (Converting factor is 1000)

Chapter 2	
Speed $= \frac{distance(s)}{time(t)}$ Average speed $= \frac{Total\ distance\ travelled(s)}{Total\ time\ taken\ (t)}$ Velocity $= \frac{displacement(s)}{time(t)}$ Acceleration $= \frac{Change\ in\ velocity}{time\ taken\ (t)}$	Acceleration $= \frac{Final\ velocity\ (v) - Initial\ velocity\ (u)}{time\ taken\ (t)}$ $v = u + at$ $s = \frac{1}{2}(u+v)t$ $s = ut + \frac{1}{2}at^2$ $v^2 - u^2 = 2as$ Where, u = initial velocity; a = acceleration; v = final velocity; s = displacement; t = time taken

Chapter 3

Force = mass x acceleration

Stopping distance = thinking distance + braking distance

Chapter 4	Chapter 5
Weight = mass x gravitational field strength Density $(\rho) = \frac{Mass\ (m)}{Volume\ (V)}$	Moment(M) = Force(F) x perpendicular distance of the force from the pivot (d) At equilibrium, Clockwise moment = Anticlockwise moment

Chapter 6 / Chapter 7	Chapter 8
Chapter 6 Extension = Stretched length – Unstretched length **Chapter 7** Pressure (P) $= \frac{Force\ (F)}{Area(A)}$ Pressure below a liquid surface $= \rho gh$ *Pressure exerted by gas in a manometer =* Atmospheric pressure + liquid pressure Boyle's Law, 　　P1 x V1 = P2 x V2, at constant pressure	Kinetic Energy (K.E.) $= \frac{1}{2} Mass\ (m)\ x\ Velocity\ (V)^2$ GPE = *mass x gravitational force x height* *Work = force (F) × displacement (s)* *Efficiency* $= \frac{useful\ energy\ output}{total\ energy\ input}$ % *Efficiency* $= \frac{useful\ energy\ output}{total\ energy\ input}$ x 100 *Power* $= \frac{Work\ done}{Tme\ taken}$ Energy = power x time

Chapter 11	Chapter 13
E = mcΔθ Where, E = thermal energy supplied m = mass of metal block c = specific heat capacity Δθ = increase in temperature of the block ($\theta_2 - \theta_1$) θ_2 = Final temperature θ_1 = Initial temperature *thermal energy = mass × specific latent heat* $$\frac{V1}{T1} = \frac{V2}{T2}$$, at constant pressure Where, V1 is the initial volume of the gas at temperature T1. V2 is the final volume of the gas at temperature T2.	***velocity = frequency × wavelength*** *V = f x λ*
	Chapter 14
	Refractive index (n) = $\frac{Sin\ i}{Sin\ r}$ *Magnification = image size/object size* *Or* *Magnification = image distance/object distance*

Chapter 16	Chapter 21
Speed of sound waves = distance / time Velocity = frequency x wavelength	Power = Voltage × Current Energy = Voltage × Current × Time Energy = Power × Time

Chapter 19	Chapter 23

Charge (Q) = Current (I) x Time (t)
 	
Current (I) = $\frac{Charge(Q)}{time(t)}$
Where,
I = current in ampere (A)
Q= charge in coulomb (C)
t = time is seconds (s)
 	
V = $\frac{E}{Q}$ or V = $\frac{W}{Q}$
Where,
V = Electromotive force of the cell in volts (V)
E = Energy given out by the source or work done in moving the charge in Joules (J)
Q = Charge in coulombs (C)
 	
R = $\frac{\rho L}{A}$
Where,
R = resistance in ohm (Ω)
ρ = resistivity in ohm metre (Ωm)
L = length in metre (m)
A = Area of cross section (m^2)
 	
R = $\frac{V}{I}$
Where,
R = the resistance in ohm (Ω)
V = the potential difference in volts (V)
I = the current in amperes (A)
 	
The combined resistance of resistors in series is
                      R = R1 + R2
 	
The combined resistance of resistors in parallel is
                  $\frac{1}{R} = \frac{1}{R1} + \frac{1}{R2}$
	Assuming that the efficiency of the transformer is ideal (100%),
	
	Power input in the primary coil = Power output in secondary coil
	           $V_p\ I_p = V_s\ I_s$
	
	Combining all the two equations we obtain the following equation
	
	        $V_s / V_p = N_s / N_p = I_p / I_s$
	Where,
	I_p is the current in the primary coil and I_s is the current in the secondary coil.
	
	Visit: <u>www.staffordeducationalservices.com</u> for additional support.

1 (f) Symbols and units of physical quantities

Quantity	Symbol	Unit
Length	l, h	km, m, cm, mm
Area	A	m^2, cm^2
Volume	V	m^3, cm^3
Weight	W	N
Mass	m, M	kg, g, mg
Time	t	h, min, s, ms
Density	ρ	g/cm^3, kg/m^3
Speed	u, v	km/h, m/s, cm/s
Acceleration	a	m/s^2
Acceleration of free fall	g	m/s^2
Force	$F, P...$	N
Moment of force		Nm
Work done	W, E	J, kWh
Energy	E	J
Power	P	W
Pressure	p, P	Pa, N/m^2
Atmospheric pressure		use of millibar
Temperature	θ, t, T	°C
Heat capacity	C	J/°C
Specific heat capacity	c	J/(kg°C), J/(g°C)
Latent heat	L	J
Specific latent heat	l	J/kg, J/g
Frequency	f	Hz
Wavelength	λ	m, cm
Focal length	f	m, cm
Angle of incidence	i	degree (°)
Angles of reflection, refraction	r	degree (°)
Critical angle	c	degree (°)
Potential difference or voltage	V	V, mV
Current	I	A, mA
Charge		C
e.m.f.	E	V
Resistance	R	Ω

Visit: www.staffordeducationalservices.com for additional support.

Printed in Great
Britain
by Amazon